Studies in Computational Intelligence

Volume 877

The series "Studies in Computational Intelligence" (SCI) publishes new developments and advances in the various areas of computational intelligence—quickly and with a high quality. The intent is to cover the theory, applications, and design methods of computational intelligence, as embedded in the fields of engineering, computer science, physics and life sciences, as well as the methodologies behind them. The series contains monographs, lecture notes and edited volumes in computational intelligence spanning the areas of neural networks, connectionist systems, genetic algorithms, evolutionary computation, artificial intelligence, cellular automata, self-organizing systems, soft computing, fuzzy systems, and hybrid intelligent systems. Of particular value to both the contributors and the readership are the short publication timeframe and the world-wide distribution, which enable both wide and rapid dissemination of research output.

The books of this series are submitted to indexing to Web of Science, EI-Compendex, DBLP, SCOPUS, Google Scholar and Springerlink.

More information about this series at http://www.springer.com/series/7092

Mohammad Shehab

Artificial Intelligence in Diffusion MRI

Enhanced Cuckoo Search Algorithm with Metaheuristic Components for Extracting the Maxima of the Orientation Distribution Function

Springer

Mohammad Shehab
Computer Science\Artificial
Intelligence Department
Aqaba University of Technology
Aqaba, Jordan

ISSN 1860-949X ISSN 1860-9503 (electronic)
Studies in Computational Intelligence
ISBN 978-3-030-36082-5 ISBN 978-3-030-36083-2 (eBook)
https://doi.org/10.1007/978-3-030-36083-2

This Springer imprint is published by the registered company Springer Nature Switzerland AG
The registered company address is: Gewerbestrasse 11, 6330 Cham, Switzerland

About This Book

The Diffusion-Weighted Magnetic Resonance Imaging (DW-MRI) is a promising method for non-invasive investigation of anatomical connectivity in the human brain. The raw data acquired from the MRI scanner may not be directly usable by the medical specialists. The initial representation of the MRI data is a huge group of fibers. Q-Ball Imaging (QBI) is a diffusion MRI reconstruction technique which has been proven very successful in resolving multiple intravoxel fiber orientations in MRI (i.e., fiber crossing) based on the standard computation of the Orientation Distribution Function (ODF), which is a three-dimensional spherical function founded to detect the dominant fiber orientations in the underlying volume of a pixel (voxel). However, ODF suffers from a lack of local estimation accuracy along the path. This dissertation presents a new method to solve the ODF problem through adapting one of the metaheuristic algorithms, namely, Cuckoo Search Algorithm (CSA). The adaptation is tested using synthetic data. The results showed some shortcomings in the convergence rate and local exploitation. These limitations are then addressed by enhancing with known metaheuristic components. Three successive enhancement versions are proposed which are incremental improvements over the preceding version: (i) Modified Cuckoo Search Algorithm (MCSA); (ii) hybridize the MCSA with components of the bat algorithm (CSBA), and (iii) hybridize the CSBA with the hill climbing (CSAHC). The experiments were conducted through two steps: The first step involved comparing against five other methods using 13 benchmark functions of various categories. This is to test each version before using it for the ODF. The second step compares against five other methods using synthetic data to select the best version for the real data. The CSAHC-ODF was applied to the real data and achieved reliable results, especially in the complicated regions in the human brain. In other words, the ODFs reconstructed by CSAHC-ODF are sharper and more accurate ODFs than the original image and extracts more accurate maxima.

Contents

Abbreviations

BA	Bat Algorithm
BCS	Binary Cuckoo Search
BCSA	Basic Cuckoo Search Algorithm
CSA	Cuckoo Search Algorithm
CSA-ODF	Cuckoo Search Algorithm for Orientation Distribution Function
CSBA	Hybridization MCSA with Bat algorithm
CSAHC	Hybridization CSBA with Hill Climbing
CSAHC-ODF	CSAHC for Orientation Distribution Function
CSBA-ODF	CSBA for Orientation Distribution Function
DICOM	Digital Imaging and Communications in Medicine
DSCA	Discrete Cuckoo Search
DSI	Diffusion Spectrum Imaging
DTI	Diffusion Tensor Imaging
DWI	Diffusion-Weighted Imaging
FA	Fractional Anisotropy
FRT	Funk-Radon Transform
GA	Genetic Algorithm
HARDI	High Angular Resolution Diffusion Imaging
HC	Hill Climbing
HS	Harmony Search
ICSA	Improved CSA
IPPT	Institut Perubatan dan Pergigian Termaju
MCSA	Modified Cuckoo Search Algorithm
MCSA-ODF	Modified Cuckoo Search Algorithm for Orientation Distribution Function
MD	Mean Deviation
MRI	Magnetic Resonance Imaging
NIfTI	Neuroimaging Informatics Technology Initiative
ODF	Orientation Distribution Function
PAS	Persistent Angular Structure

PDF	Probability Density Function
PGSE	Pulsed Gradient Spin Echo
PSO	Particle Swarm Optimization
QBI	Q-Ball Imaging
RBF	Radial Basis Function
RSSH	Real Symmetric Spherical Harmonics
SH	Spherical Harmonic
SMES	Super Conducting Magnetic Energy Storage
SNR	Signal-to-Noise Ratio
t	Tournament Size
TSE	Turbo Spin Echo

List of Figures

List of Tables

Chapter 1
Introduction of Diffusion MRI and Cuckoo Search Algorithm

1.1 Background

The brain is the most complex organ in the human body because it consists of about 100 billion neurons and one million billion (10^{15}) interconnections (Azevedo et al. 2009). This organ is the control for the sensorimotor such as walking and breathing, cognitive functions such as talking, reasoning, memory and more complex functions such as emotions and feelings. The brain is also a subject of many diseases that need surgery, which could result in either deterioration of the cited functions or even in permanent disability. Medical imaging, especially Magnetic Resonance Imaging (MRI), helps mapping the anatomical and functional aspects of the brain, considered as the substratum of the different functions. In the first part of this section, a brief overview of the diffusion MRI is presented, followed by an overview of the Cuckoo Search Algorithm (CSA). The justifications for using the CSA to solve the ODF problem are given at the end of this section.

1.1.1 Diffusion MRI

The central nervous system is made up of neurones. A neurone is constituted of two types of tissue, namely the gray matter where cell bodies of neurons reside in the brain and spinal cord, and the axone forming the white matter, which are extensions of neural cells as shown in Fig. 1.1. Integration of the neural processes in the human brain is realized through interconnections that exist between different neural centers (cortical centers). This connectivity is altered by some pathologies or trauma, resulting in deterioration of sensorimotor and cognitive capabilities. For this reason, it is of high importance to probe this connectivity and evaluate its integrity.

Diffusion MRI measures the diffusion of water molecules along a set of directions (Le Bihan et al. 1986). Diffusion MRI is a noninvasive method based on the Brownian (random) motion of water molecules constrained by neuronal tissues in vivo. During

M. Shehab, *Artificial Intelligence in Diffusion MRI*, Studies in Computational Intelligence 877, https://doi.org/10.1007/978-3-030-36083-2_1

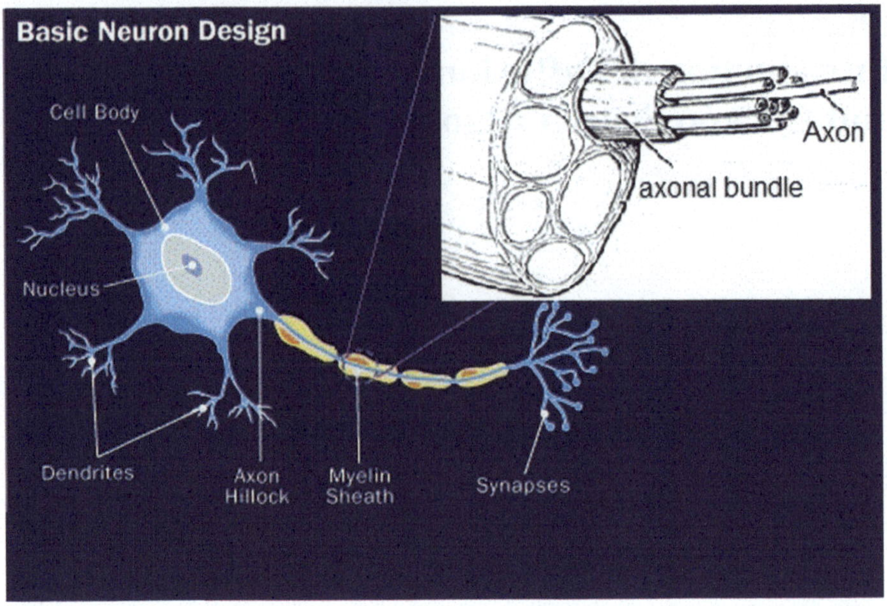

Fig. 1.1 Structure of a typical neuron (Craig and Robynne 2001)

the early 2000s, various research groups have been proposed to build structural connection matrices from diffusion MRI fiber tracking through employing different diffusion acquisition techniques (Thottakara et al. 2006; Iturria-Medina et al. 2007; Gong et al. 2009).

The most commonly used diffusion MRI technique is diffusion weighted imaging (DWI) (Taylor and Bushell 1985; Le Bihan and Breton 1985). DWI is a non-invasive technique which may useful for detecting and characterizing the pathological and non-pathological features of living tissue. It is also very sensitive to water movements (molecular Brownian motion) (Yan 2015). DWI is able to determine areas affected and healthy areas, but it can not provide details about the fibers directions in the MRI diffusion (Gass et al. 2004).

Diffusion Tensor Imaging (DTI) is the answer to the limitations of DWI through using tensors, which are mathematical tools capable of describing diffusion in different space directions, because diffusion is in most of the cases anisotropic[1] (Basser et al. 1994). DTI is a powerful technique to evaluate the major white matter fibre bundles. It also has a positive impact on disease prognosis, neurosurgical resection, and the preservation of brain function (Romano et al. 2009). Nevertheless, some challenges negatively affect the accuracy and validity of results of this technique. For example, in the areas of complex intravoxel, the DTI technique fails to char-

[1]Anisotropic is highly structured and typically have different diffusion coefficients along different directions (Tariq et al. 2016).

acterize the multiple fibre directions accurately in reconstructing fanning, crossing, kissing and bending fibres (Kuhnt et al. 2013b; Pujol et al. 2015).

Therefore, High Angular Resolution Diffusion Imaging (HARDI) was introduced by Tuch et al. (1999) to overcome the limitations of the DTI. HARDI requires a very large number of DWI. So, it provides more diffusion information (Alaya et al. 2017). Examples of the HARDI, Diffusion Spectrum Imaging (DSI) is developed to image complex distributions of intravoxel fibre orientation that is capable of mapping fibre architectures by imaging the 3D spectra of water molecules' displacement (Wedeen et al. 2000). Therefore, DSI is considered a model-free method (i.e., It does not assume a particular diffusion model, such as tensor model and multiple-tensor model) (Zucchelli et al. 2014; Sperl et al. 2017). However, the number of collected points from the q-space of DSI is more than ten times greater than DTI, this leads to long acquisition times and difficult to implement in a clinical application (Bilgic et al. 2012; Young et al. 2017). Thus, Q-Ball Imaging (QBI) was introduced by Tuch (2004b) to overcome the drawbacks of the DTI in the intravoxel heterogeneities of fibre orientation and the DSI problem in the acquisition time. QBI computes the Orientation Density Function (ODF) which is the radial projection of the Probability Density Function (PDF) modeled as a spherical function able to represent crossing fibres (Pontabry et al. 2013; Fan et al. 2016).

Basically, the ODF is a criterion to determine the fibres' directions within a certain voxel. The fibre's path can be extracted in some directions corresponding to the highest orientation likelihood. In other words, an ODF may be considered as a deformed sphere whose radius in a given direction is proportional to the sum of values of the diffusion PDF in that direction (Hagmann et al. 2006). To further ease visualization, the surface of the ODF is colour coded according to a diffusion direction. Where x, y, and z coordinates refer to red, green, and blue, respectively (Topgaard 2017).

In general, optimization algorithms seeking to find the best effective cost or achieving the highest possible level of interest (maximizing or minimizing) by systematically choosing the values of decision variables from a feasible set while satisfying a given set of constraints (Rao and Patel 2013). Optimization algorithms proved its worth by achieving best results in different problems. For example, large tourism companies use optimization models to schedule routes and places to visit to achieve the maximum profit (Qiu et al. 2002). Shipping companies use optimization models to determine the best ship speed to reach its destination with the lowest fuel cost (Mansouri et al. 2015). Routers use optimization models to select the best path to forward data packets (Khedr et al. 2015).

1.1.2 Cuckoo Search Algorithm

The Cuckoo Search Algorithm (CSA) introduced by Yang and Deb (2009), it is inspired by the behavior of cuckoo breeding. CSA is based on the obligate brood parasitic behavior found in several cuckoo species combined with the *Lévy* flight behavior discovered in some birds and fruit flies. This algorithm has attracted attention since 2009 (Shehab et al. 2017b).

Yang and Deb used a specific and simple representation for implementing CSA with each egg representing a solution. As each cuckoo lays only one egg, it also represents one solution. This representation aims to increase the diversity of new and probably good cuckoos (solutions) and to replace the unfit solutions. CSA can be complicated by using multiple eggs in each nest to represent a set of solutions (Shehab et al. 2017c, 2019a).

The breeding behavior of a cuckoo is aggressive in nature, which inspires the optimization algorithms. Brood parasitism is a primary mechanism of a cuckoo, this bird lays eggs in the host's nest and carefully matches its eggs through mimicking the pattern and color (Rajabioun 2011). In the case the host recognizes the cuckoo egg in its own nest, the host will either throw the egg out or simply leave its nest and build a new one. Therefore, a cuckoo must be accurate in its mimicry of the host eggs (Shehab et al. 2019b). By contrast, the host tries to improve its skills of determining the parasitic egg, which is called the struggle for survival (Shehab et al. 2018a).

The CSA provides solutions in a reasonable computational time and cost (Feng et al. 2014; Yang 2014). This algorithm is more easily implemented compared to other population metaheuristic algorithms such as Harmony Search (HS) and Genetic Algorithm (GA) because it has two parameters the probability of being discovered by the host bird P_α and the population n host nests. Therefore, CSA is one of the simplest algorithms and considered an effective search approach for solving complex optimization problems (Shehab et al. 2017a). CSA also has two main operations: a random search based on the probability of the host bird to detect alien eggs and direct search based on *Lévy* flights (Kamalakannan et al. 2014). In other words, CSA balances between the global exploration and the deep exploitation in the search space. As such, CSA has been successfully applied to solve a broad range of real-world optimization problems (Li and Yin 2016).

1.1.3 Why CSA Has Been Chosen to Solve the ODF

Swarm-based algorithms are optimization algorithms containing some algorithms inspired by nature such as artificial bee colony (ABC) (Karaboga 2005), particle swarm optimization (PSO) (James and Russell 1995), firefly algorithm (FA) (Yang and Algorithms 2008), and CSA. These algorithms have many advantages over conventional algorithms. They combine rules and randomness to imitate some natural phenomena (Siddique and Adeli 2015). In addition, they are efficient, highly adaptable, and tend to be flexible to implement. In other words, the features of these algorithms make it possible to use them in a wide range of problems in varied applications (Yang 2015; Shehab and Khader 2018).

Based on the above, it can use any swarm-based algorithm to solve the ODF problem. However, during our research, we found that the CSA is the best choice to solve the ODF problem. For instance, it has fewer parameters, and one of the most powerful features of CSA is the use of *Lévy* flights to generate new candidate solutions (Shehab et al. 2018b). Thus, the CSA is a suitable algorithm for the ODF problem. Furthermore, Precision in the mechanism of selection the solutions in the CSA is suitable for mechanism of the ODF to select the optimal point. In the CSA, the search is based on the *Lévy* flights, which moves between the solutions to determine a new solution and compares it with the random solution. It also uses the abandon of the worst solution (i.e., probability $P\alpha$). Therefore, the chosen solution should have a high fitness value to exceed these conditions. On the other hand, the ODF is obtained from diffusion weighted signals measured using the QBI. The spherical harmonics have been used to represent the ODF where each point in the sphere is represented through angle (θ, ϕ). Detecting the maxima of the ODF requires navigating between these points and extracting the maxima carefully. This process is available in the CSA in the selection solution. Due to these reasons, the CSA was selected to detect the ODF maxima.

1.2 Motivation and Problem Statement

The human brain is considered the center for neurons. As mentioned previously, this organ is responsible for sensorimotor and cognitive functions which is behind all aspects of human life. Brain pathologies or trauma lead most of the time to permanent disability and life quality deterioration, which require taking all kind of precautions when intervening on this organ.

Fibre tracking is a non-invasive tool that is used to visualize and measure the pathways of white matter in the human brain initiated using the DTI (Wedeen et al. 1995; Weiss et al. 2015). Nevertheless, some challenges negatively affect the accuracy and validity of the results of this tool. For example, in the areas of complex intravoxel, the DTI technique fails to characterize the multiple fibre directions accurately in reconstructing crossing or kissing fibres (Wedeen et al. 2008; Kuhnt et al. 2013a; Shehab et al. 2016). Consequently, DTI fails to describe the diffusion process accurately, leads to influencing the efficiency of the fiber tracking algorithms.

The limitations of DTI have been overcome by introducing the QBI technique, which it is a variant of DWI that is sensitive to intravoxel heterogeneities in diffusion directions caused by crossing fibre tracts. QBI is a HARDI technique which has been proven very successful in resolving multiple fibre crossings and branchings in multiple intravoxel fibres using standard computation of ODF by directly sampling the diffusion signal on a spherical shell in diffusion space (Tomána et al. 2007). However, ODF still has a limitation in determining fiber directions which may be corrupted by neighbor directions (Assemlal et al. 2009).

In other words, ODF suffers from a lack of local estimation accuracy along the path (Çetingül et al. 2011; Daducci et al. 2015). Deterministic methods have been used by Thomas et al. (2014) to deal with such limitations. However, these methods suffer from local noise-induced disturbances which are additively accumulated along the track propagation, and lack of connectivity information between regions of the brain

(Jones and Pierpaoli 2005). Therefore, probabilistic methods have been employed by Vorburger (2012) to increase the accuracy of the local estimation along the path, but it also suffers from very long processing times, preventing their use in the clinical field (Parker 2014; Abualigah et al. 2017b). So, it is necessary to find alternative ways to address these limitations.

The optimization algorithm is a procedure which is executed iteratively by comparing different solutions till a satisfactory or an optimum solution is found (Deb 2012). Wherefore, optimization will be conducted by employing the basic CSA (BCSA), which is considered a novel population-based stochastic global search metaheuristic algorithm (Zhao et al. 2012). According to the best knowledge available, BCSA efficiently handles the exploitation and exploration (Cuevas and Reyna-Orta 2014). However, existing works have not investigated the BCSA in the context of the ODF in general. The focus of this dissertation is to adapt BCSA for ODF to extract all the fibres directions, then determine the maxima as a subset of the stationary points of the ODF with the possibility of controlling the convergence (balancing between slow and premature convergence) with increasing the quality and the efficiency of fiber direction. After that, enhancing its efficiency by combining components from metaheuristics.

1.3 Research Objectives

The aim of this dissertation is not only to adapt the cuckoo search algorithm for orientation distribution function but also to introduce new versions of the cuckoo search algorithm for detecting the maximum of the ODF. Thus, new alternatives for solving ODF problems are provided.

The main objectives of the research presented in this dissertation are as follows:

1. To adapt the basic cuckoo search algorithm for orientation distribution function.
2. To enhance the adapted BCSA by injecting heuristic-based techniques to improve the quality of the solution.
3. To assess the usefulness of the best version of BCSA for applying it on human brain data.

1.4 Research Contributions

The key contributions of this dissertation to the literature are:

1. Adapting the BCSA for ODF and extracting the ODF maximum, henceforth called Cuckoo Search Algorithm for Orientation Distribution Function (CSA-ODF).
2. Introducing three versions of the BCSA with other metaheuristic components. These new versions can be described as follows:

(a) **Modified Cuckoo Search Algorithm**:
The Modified Cuckoo Search Algorithm (MCSA) is based on swapping the random selection (original) with tournament selection (proposed). Thus, the probability of better results is increased, through giving the solution that has the high fitness value a chance to be selected. That leads to increasing the quality of the solutions, increasing the diversity, and enhancing the convergence. MCSA is used to solve the ODF problem, namely, MCSA-ODF.

(b) **Hybridization of modified cuckoo search algorithm with parts of the bat algorithm components**
Hybridization of MCSA and some components of BA, namely, CSBA. This version starts with the MCSA to set up the population of host nests then looking for a new solution through the components of the BA. In such a way, CSBA focuses on the exploitation which leads to improving the convergence (avoiding the slow convergence) of the MCSA. The proposed version is used to solve the ODF problem, namely, CSBA-ODF.

(c) **Hybridization of cuckoo search algorithm with Hill Climbing**:
Hybridizing CSBA with local search -based algorithm (i.e., hill climbing (HC)), namely, CSAHC. This version proposed to overcome the drawbacks of the CSBA. In other words, the CSBA becomes more exploitative very fast and may stack at a local optimum. Thus, CSAHC used the mechanism of the HC as a new operator to increase its ability to find the local optimal solution in the search space. This version is used for solving the ODF problem which called CSAHC-ODF.

3. The applicability of the best version of the CSA-ODF versions is comprehensively assessed using real data of the human brain acquired from Institut Perubatan dan Pergigian Termaju -Universiti Sains Malaysia (IPPT-USM).

1.5 Scope of the Research

This research focuses on enhancing the performance of the BCSA to extract the ODF maxima. There are three versions of BCAS: (i) MCSA, that replacing the current selection (i.e., random) with another selection scheme (i.e., tournament). (ii) CSBA, that hybridized the MCSA with components of the BA. (iii) CSAHC, that hybridized CSBA with HC.

The research is limited to use set of benchmark functions (i.e., unimodal and multimodal problems) to evaluate each version of BCSA (i.e., MCSA, CSBA, and CSAHC), as well as it is limited to synthetic data to assess the effectiveness of the proposed versions of the BCSA with ODF (i.e., CSA-ODF, MCSA-ODF, CSBA-ODF, and CSAHC-ODF) with performance comparison of these versions to choose the best version. Regarding the best version, this research is limited to real human brain data to evaluate the efficiency and accuracy of the best version through extracting the ODF maxima with high quality and acceptable time.

1.6 Overview of Methodology

This section provides a brief discussion on the methodology, described fully in Chap. 4, to achieve the research objectives.

In Fig. 1.2, stage 1 shows that after preparing the data, a new ODF maxima search approach using the BCSA has been proposed to extract all the ODF maxima (i.e., CSA-ODF) which achieve the first objective, the synthetic data is used to evaluate the CSA-ODF. To improve the performance of the CSA-ODF, the approach is worthy of further research in order to achieve the second objective.

Stage 2 shows the second objective which contains three techniques of effective components from metaheuristics were incorporated into BCSA. The first enhancement is the selection schemes concept from the tournament selection scheme used to enhance BCSA results by giving the opportunity for the best solution to be chosen (i.e., MCSA). The second enhancement is the hybridization which combines MCSA and a part from the BA (i.e., CSBA). This leads to focusing on the exploitation search which is characterized by BA thereby overcoming the slow convergence suffered by the MCSA. The third enhancement is the local search-based algorithm hybridization that combines CSBA and HC operator (i.e., CSAHC) to improve the local exploitation of the CSBA. For evaluation each version of BCSA (without ODF; MCSA, CSBA, and CSAHC) a set of benchmark functions are used, while the synthetic data is used to evaluate the performance of each version of the CSA-ODF (with ODF; MCSA-ODF, CSBA-ODF, and CSAHC-ODF).

For the third objective, the best version which will be chosen after comparing all versions will be evaluated by using human brain data collected from IPPT-USM, can be seen in Fig. 1.2- stage 3.

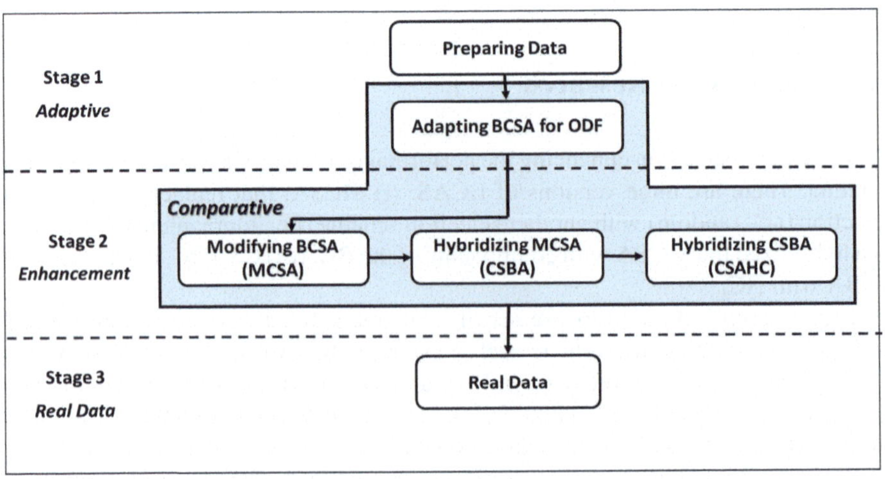

Fig. 1.2 Research methodology

1.7 Overview of the Dissertation

This dissertation includes nine chapters organized as follows: Chap. 2 provides an overview of the diffusion MRI. It also surveys some previous methods that tackled the ODF problems.

Chapter 3 discusses the cuckoo search algorithm in details; procedure, growth, and variants of CSA. The methodology is proposed and described in details in Chap. 4.

Chapter 5 is divided into two main parts. The first part, introduces the steps of pre-processing such as generate the synthetic data. The second part, describes the adaptive BCSA to solve ODF (CSA-ODF), then evaluate fiber detection success through using the synthetic data.

Chapters 6 and 7 illustrate the Modification (MCSA-ODF) and two types of hybridization (CSBA-ODF, CSAHC-ODF), consecutively. The experiments and results with detailed analysis of studying are presented at each chapter. It should be noted that the experiments were done through two stages of evaluation. (1) Applied set of benchmark functions optimization, followed by (2) Applied synthetic data.

Thereafter, applied The best version of CSA-ODF versions on the human brain's data is described in Chapter 8.

Finally, in Chap. 9 a summary of claimed results and future possibilities are provided. It is hoped that the reader will be challenged to prove or disprove the claims made throughout this dissertation, and extend the research in new directions.

References

Abualigah, L. M., Sawaie, A. M., Khader, A. T., Rashaideh, H., Al-Betar, M. A., & Shehab. M. (2017b). β-hill climbing technique for the text document clustering. *New Trends in Information Technology, 60.*

Alaya, I. B., Jribi, M., Ghorbel, F., Sappey-Marinier, D., & Kraiem, T (2017). Fast and accurate estimation of the hardi signal in diffusion mri using a nearest-neighbor interpolation approach. *IRBM, 38*(3), 156–166.

Assemlal, H.-E., Tschumperlé, D., & Brun, L. (2009). Efficient and robust computation of pdf features from diffusion mr signal. *Medical Image Analysis, 13*(5), 715–729.

Azevedo, F. A. C., Carvalho, L. R. B., Grinberg, L. T., Farfel, J. M., Ferretti, R. E. L., Leite, R. E. P., Lent, R., & Herculano-Houzel, S et al. (2009). Equal numbers of neuronal and nonneuronal cells make the human brain an isometrically scaled-up primate brain. *Journal of Comparative Neurology, 513*(5), 532–541.

Basser, P. J., Mattiello, J., & LeBihan, D. (1994) Mr diffusion tensor spectroscopy and imaging. *Biophysical Journal, 66*(1), 259.

Bilgic, B., Setsompop, K., Cohen-Adad, J., Yendiki, A., Wald, L. L., & Adalsteinsson, E. (2012). Accelerated diffusion spectrum imaging with compressed sensing using adaptive dictionaries. *Magnetic Resonance in Medicine, 68*(6), 1747–1754.

Çetingül, H.E., Plank, G., Trayanova, N. A., & Vidal, R. (2011). Estimation of local orientations in fibrous structures with applications to the purkinje system. *IEEE Transactions on Biomedical Engineering, 58*(6), 1762–1772.

Craig, F., & Robynne, B. (2001). How your brain works. https://science.howstuffworks.com/life/inside-the-mind/human-brain/brain.htm.

Cuevas, E., & Reyna-Orta, A. (2014). A cuckoo search algorithm for multimodal optimization. *The Scientific World Journal*.

Daducci, A., Dal Palù, A., Lemkaddem, A., & Thiran, J.-P. (2015). Commit: convex optimization modeling for microstructure informed tractography. *IEEE Transactions on Medical Imaging*, *34*(1), 246–257.

Deb, K. (2012). *Optimization for engineering design: Algorithms and examples*. PHI Learning Pvt. Ltd.

Denis, L. B., & Breton, E. (1985). Imagerie de diffusion in-vivo par résonance magnétique nucléaire. *Comptes-Rendus de l'Académie des Sciences*, *93*(5), 27–34.

Fan, Q., Witzel, T., Nummenmaa, A., Van Dijk, K. R. A., Van Horn, J. D., Drews, M. K., Somerville, L. H., Sheridan, M. A., Santillana, R. M., Snyder, J. et al. (2016). Mgh–usc human connectome project datasets with ultra-high b-value diffusion mri. *Neuroimage*, *124*, 1108–1114.

Feng, Y., Jia, K., & He, Y. (2014). An improved hybrid encoding cuckoo search algorithm for 0–1 knapsack problems. *Computational Intelligence and Neuroscience*, *2014*, 1.

Gass, A., Ay, H., Szabo, K., & Koroshetz, W. J. (2004). Diffusion-weighted mri for the small stuff: the details of acute cerebral ischaemia. *The Lancet Neurology*, *3*(1), 39–45.

Gong, G., He, Y., Concha, L., Lebel, C., Gross, D. W., Evans, A. C., & Beaulieu, C. (2009). Mapping anatomical connectivity patterns of human cerebral cortex using in vivo diffusion tensor imaging tractography. *Cerebral Cortex*, *19*(3), 524–536.

Hagmann, P., Jonasson, L., Maeder, P., Thiran, J.-P., Van Wedeen, J., & Meuli, R. (2006). Understanding diffusion mr imaging techniques: From scalar diffusion-weighted imaging to diffusion tensor imaging and beyond 1. *Radiographics*, *26*(suppl_1), S205–S223.

Iturria-Medina, Y., Canales-Rodriguez, E. J., Melie-Garcia, L., Valdes-Hernandez, P. A., Martinez-Montes, E., Alemán-Gómez, Y., et al. (2007). Characterizing brain anatomical connections using diffusion weighted mri and graph theory. *Neuroimage*, *36*(3), 645–660.

James, K., & Russell, E. (1995). Particle swarm optimization. In *Proceedings of 1995 IEEE International Conference on Neural Networks*, pp. 1942–1948.

Jones, D. K. & Pierpaoli, C. (2005). Confidence mapping in diffusion tensor magnetic resonance imaging tractography using a bootstrap approach. *Magnetic Resonance in Medicine*, *53*(5), 1143–1149.

Kamalakannan, C., Suresh, P., Dash, S. S., & Panigrahi, B. K. (2014). *Power Electronics and Renewable Energy Systems: Proceedings of ICPERES 2014*, vol. 326. Springer.

Karaboga, D. (2005). An idea based on honey bee swarm for numerical optimization. Technical report, Technical report-tr06, Erciyes university, engineering faculty, computer engineering department.

Khedr, M. E., Zaghloul, M. S., & El-Desouky, M. I. (2015). Wireless adhoc multi access networks optimization using ospf routing protocol based on cisco devices. *International Journal of Computer Networks & Communications*, *7*(2), 59.

Kuhnt, D., Bauer, M. H. A., Egger, J., Richter, M., Kapur, T., Sommer, J., Merhof, D., & Nimsky, C. (2013a). Fiber tractography based on diffusion tensor imaging compared with high-angular-resolution diffusion imaging with compressed sensing: initial experience. *Neurosurgery*, *72*(0 1), 165.

Kuhnt, D., Bauer, M. H. A., Sommer, J., Merhof, D., & Nimsky, C. (2013b). Optic radiation fiber tractography in glioma patients based on high angular resolution diffusion imaging with compressed sensing compared with diffusion tensor imaging-initial experience. *PLoS One*, *8*(7), e70973.

Le Bihan, D., Breton, E., Lallemand, D., Grenier, P., Cabanis, E., & Laval-Jeantet, M. (1986). Mr imaging of intravoxel incoherent motions: Application to diffusion and perfusion in neurologic disorders. *Radiology*, *161*(2), 401–407.

Li, X., & Yin, M. (2016). A particle swarm inspired cuckoo search algorithm for real parameter optimization. *Soft Computing*, *20*(4), 1389–1413.

Mansouri, S. A., Lee, H., & Aluko, O. (2015). Multi-objective decision support to enhance environmental sustainability in maritime shipping: a review and future directions. *Transportation Research Part E: Logistics and Transportation Review*, *78*, 3–18.

Parker, G. J. M. (2014). Analysis of mr diffusion weighted images. *The British Journal of Radiology*.

Pontabry, J., Rousseau, F., Oubel, E., Studholme, C., Koob, M., & Dietemann, J.-L. (2013). Probabilistic tractography using q-ball imaging and particle filtering: application to adult and in-utero fetal brain studies. *Medical Image Analysis*, *17*(3), 297–310.

Pujol, S., Wells, W., Pierpaoli, C., Brun, C., Gee, J., Cheng, G., et al. (2015). The dti challenge: Toward standardized evaluation of diffusion tensor imaging tractography for neurosurgery. *Journal of Neuroimaging*, *25*(6), 875–882.

Qiu, L., Hsu, W.-J., Huang, S.-Y., & Wang, H. (2002). Scheduling and routing algorithms for agvs: A survey. *International Journal of Production Research*, *40*(3), 745–760.

Rajabioun, R. (2011). Cuckoo optimization algorithm. *Applied Soft Computing*, *11*(8), 5508–5518.

Rao, R. V., & Patel, V. (2013). Multi-objective optimization of heat exchangers using a modified teaching-learning-based optimization algorithm. *Applied Mathematical Modelling*, *37*(3), 1147–1162.

Romano, A., Dandrea, G., Minniti, G., Mastronardi, L., Ferrante, L., Fantozzi, L. M., et al. (2009). Pre-surgical planning and mr-tractography utility in brain tumour resection. *European Radiology*, *19*(12), 2798.

Shehab, M., & Khader, A. T. (2018). *Modified cuckoo search algorithm using a new selection scheme for unconstrained optimization problems*, *14*, 1.

Shehab, M., Daoud, M. Sh., AlMimi, H. M., Abualigah, L. M., & Khader, A. T. (2019a). Hybridizing cuckoo search algorithm for extracting the odf maxima in spherical harmonic representation. *International Journal of Bio-Inspired Computation*, (in press).

Shehab, M., Khader, A. T., & Al-Betar, M. A. (2016). New selection schemes for particle swarm optimization. *IEEJ Transactions on Electronics, Information and Systems*, *136*(12), 1706–1711. https://doi.org/10.1541/ieejeiss.136.1706.

Shehab, M., Khader, A. T., & Al-Betar, M. A. (2017a). A survey on applications and variants of the cuckoo search algorithm. *Applied Soft Computing*.

Shehab, M., Khader, A. T., & Alia, M. A. (2019b). Enhancing cuckoo search algorithm by using reinforcement learning for constrained engineering optimization problems. In *2019 IEEE Jordan International Joint Conference on Electrical Engineering and Information Technology (JEEIT)*, pp. 812–816. IEEE.

Shehab, M., Khader, A. T., & Laouchedi, M. (2017c). Modified cuckoo search algorithm for solving global optimization problems. In *International Conference of Reliable Information and Communication Technology*, pp. 561–570. Springer.

Shehab, M., Khader, A. T., & Laouchedi, M. (2018a). A hybrid method based on cuckoo search algorithm for global optimization problems. *Journal of ICT*, *17*(3), 469–491.

Shehab, M., Khader, A. T., Al-Betar, M. A., & Abualigah, L. M. (2017b). Hybridizing cuckoo search algorithm with hill climbing for numerical optimization problems. In *Information Technology (ICIT), 2017 8th International Conference on*, pp. 36–43. IEEE.

Shehab, M., Khader, A. T., Laouchedi, M., & Alomari, O. A. (2018b). Hybridizing cuckoo search algorithm with bat algorithm for global numerical optimization. *The Journal of Supercomputing*, 1–28.

Siddique, N., & Adeli, H. (2015). Nature inspired computing: An overview and some future directions. *Cognitive Computation*, *7*(6), 706–714.

Sperl, J. I., Sprenger, T., Tan, Ek. T., Menzel, M. I., Hardy, C. J., & Marinelli, L. (2017) Model-based denoising in diffusion-weighted imaging using generalized spherical deconvolution. *Magnetic Resonance in Medicine*.

Tariq, M., Schneider, T., Alexander, D. C., Claudia, A. G. (2016). Wheeler-Kingshott, and Hui Zhang. Bingham–noddi: Mapping anisotropic orientation dispersion of neurites using diffusion mri. *NeuroImage*, *133*, 207–223.

Taylor, D. G., & Bushell, M. C. (1985). The spatial mapping of translational diffusion coefficients by the nmr imaging technique. *Physics in Medicine and Biology, 30*(4), 345.

Thomas, C., Frank, Q. Y., Irfanoglu, M. O., Modi, P., Saleem, K. S., Leopold, D. A., & Pierpaoli, C. (2014). Anatomical accuracy of brain connections derived from diffusion mri tractography is inherently limited. *Proceedings of the National Academy of Sciences, 111*(46), 16574–16579.

Thottakara, P., Lazar, M., Johnson, S. C., & Alexander, A. L. (2006). Application of brodmann's area templates for roi selection in white matter tractography studies. *Neuroimage, 29*(3), 868–878.

Tomána, H., Tornaib, R., & Zicharc, M. (2007). Complex fiber visualization. *Annales Mathematicae et Informaticae* (pp. 103–109)., volume 34 Institute of Mathematics and Computer Science: Eszterházy Károly College.

Topgaard, D. (2017). Multidimensional diffusion mri. *Journal of Magnetic Resonance, 275*, 98–113.

Tuch, D. S. (2004b). Q-ball imaging. *Magnetic resonance in medicine, 52*(6), 1358–1372.

Tuch, D. S., Weisskoff, R. M., Belliveau, J. W., & Wedeen, V. J. (1999). High angular resolution diffusion imaging of the human brain. In *Proceedings of the 7th Annual Meeting of ISMRM, Philadelphia*, volume 321.

Vorburger, R. S. (2012). Probabilistic techniques in diffusion weighted imaging and fiber tractography.

Wedeen, V. J., Davis, T. L., Weisskoff, R. M., Tootell, R., Rosen, B. R., & Belliveau, J. W. (1995). White matter connectivity explored by mri. (Vol. 69).

Wedeen, V. J., Reese, T. G., Tuch, D. S., Weigel, M. R., Dou, J. G., Weiskoff, R. M., & Chessler, D. (2000). Mapping fiber orientation spectra in cerebral white matter with fourier-transform diffusion mri. In *Proceedings of the 8th Annual Meeting of ISMRM, Denver*, p. 82.

Wedeen, J. Van, Wang, R. P., Schmahmann, J. D., Benner, T., Tseng, W. Y. I., Dai, G., Pandya, D. N., Hagmann, P., D'Arceuil, P., & de Crespigny, A. J. (2008). Diffusion spectrum magnetic resonance imaging (dsi) tractography of crossing fibers. *Neuroimage, 41*(4), 1267–1277.

Weiss, C., Tursunova, I., Neuschmelting, V., Lockau, H., Nettekoven, C., Oros-Peusquens, A.-M. (2015). Gabriele Stoffels, Anne K Rehme, Andrea Maria Faymonville, N Jon Shah, et al. Improved ntms-and dti-derived cst tractography through anatomical roi seeding on anterior pontine level compared to internal capsule. *NeuroImage: Clinical, 7*, 424–437.

Yan, L. (2015). *Operative Techniques in liver resection*. Springer.

Yang, X.-S. & Nature-Inspired Metaheuristic Algorithms. (2008). *Luniver press*. UK: Beckington.

Yang, X.-S. (2014). Cuckoo search and firefly algorithm: Overview and analysis. In *Cuckoo Search and Firefly Algorithm*, pp. 1–26. Springer.

Yang, X.-S. (2015). Nature-inspired algorithms: Success and challenges. In *Engineering and Applied Sciences Optimization*, pp. 129–143. Springer.

Yang, X.-S., & Deb, S. (2009). Cuckoo search via lévy flights. In *Nature & Biologically Inspired Computing, 2009. NaBIC 2009. World Congress on*, pp. 210–214. IEEE.

Young, R. J., Tan, Ek. T., Peck, K. K., Jenabi, M., Karimi, S., Brennan, N., Rubel, J., Lyo, J., Shi, W., & Zhang, Z. et al. (2017). Comparison of compressed sensing diffusion spectrum imaging and diffusion tensor imaging in patients with intracranial masses. *Magnetic Resonance Imaging, 36*, 24–31, 2017.

Zhao, X. C., Huang, P. Y., Liu, T. T., & Li, X. M. (2012). A hybrid clonal selection algorithm for quality of service-aware web service selection problem. *Int J Innov Comput Inf Control, 8*(12), 8527–8544.

Zucchelli, M., Garyfallidis, E., Paquette, M., Merlet, S., Menegaz, G., & Descoteaux, M. (2014). Comparison between discrete and continuous propagator indices from cartesian q-space dsi sampling. In *ISMRM: International Society for Magnetic Resonance in Medicine*, p. 4294.

Chapter 2
Background of Diffusion MRI

2.1 Introduction

Decoding human brain structures and their interconnecting trajectories are very exciting research areas since they have numerous applications in the clinical diagnosis and management of brain disorders. For that, a lot of variety of invasive tools have been introduced to study brain white matter structural connectivity and configuration. Nevertheless, these tools pose a risk to human life. Therefore, there must be alternatives to avoid this risk.

This chapter highlighting the diffusion MRI which it is the only non-invasive method (Barmpoutis and Vemuri 2009). It is mainly organized into four sections. Diffusion MRI is discussed in Sect. 2.2. In Sect. 2.3, the diffusion in the human brain is presented in details. The fundamental of MRI techniques are illustrated in Sect. 2.4. Section 2.5, presents an overview of the previous related works. Finally, the summary is shown in Sect. 2.6.

2.2 Diffusion MRI

MRI is a complicated machine that uses pulses of radio wave energy and magnetic field to scan the interior parts of the human body (Ogawa et al. 1990). Figure 2.1 shows the MRI machine, the output is an image volume in which slices can be viewed at any location or direction. MRI is used to detect various conditions, such as injury, bleeding, tumors, infection, and blood vessel diseases (Johansen-Berg and Behrens 2013).

Diffusion MRI is an extension of conventional MRI that focuses on mapping the diffusion process of molecules as a cause of Brownian motion. Brownian motion, which was discovered by Einstein (1956), is the random motion of molecules resulting from thermal energy, see Fig. 2.2. In a glass of water, the motion of water

M. Shehab, *Artificial Intelligence in Diffusion MRI*, Studies in Computational Intelligence 877, https://doi.org/10.1007/978-3-030-36083-2_2

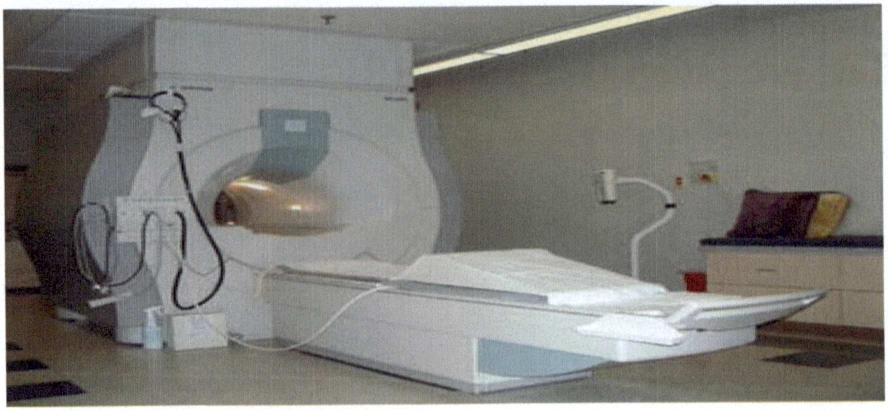

Fig. 2.1 MRI machine

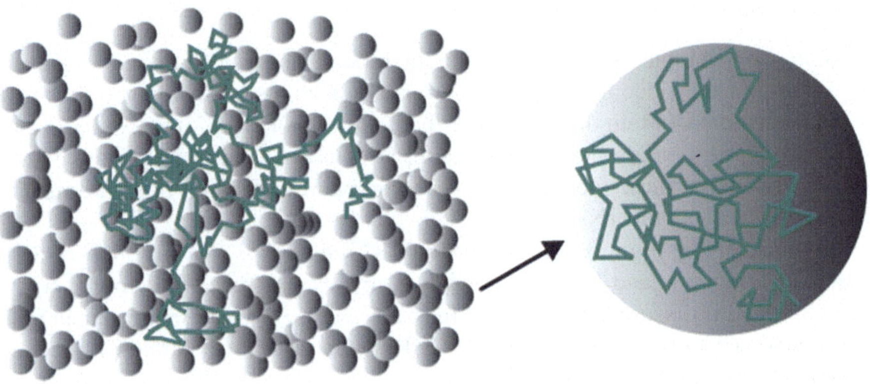

Fig. 2.2 Brownian motion of extracellular water molecules (left). Diffusion inside the cell (right)

molecules is completely random and limited only by the boundaries of the container. This motion is appropriately described by a displacement distribution that describes the proportion of molecules that undergo displacement in a specific direction. The diffusion of a molecule in a homogeneous medium is also well described in a Gaussian distribution (Shehab et al. 2016). However, the neuronal tissue has a fibrillar structure; thus, the movement of the molecules is more significantly hindered in the direction perpendicular to the axonal orientation than parallel to it. Consequently, a distribution revealing the pathway of the fiber is obtained. Diffusion MRI can depict this motion by adding a sequence to the normal MRI procedure. This procedure involves the application of a gradually increasing gradient field in the same direction as the radio frequency pulse introduces an additional phase shift to the magnetic moment proportional to the molecular displacement along that gradient direction (Hagmann et al. 2006). The large number of molecules that has moved in the gradient direction during the acquisition will result in a large phase shift. This shift subsequently leads to a loss of signal, which is used to measure the diffusion, for more details see Sect. 2.3.1.

2.3 Diffusion MRI in the Human Brain

Diffusion MRI is used for imaging the brain's white matter, which is made of microscopic fiber, glial cells, axons, and neurons (Assaf and Pasternak 2008). According to the Brownian motion of water, the white matter movement follows the Gaussian distribution, which is a continuous function for an approximation of the exact binomial distribution of events. Therefore, the molecules have restricted diffusion, and this effect is observed only with sufficient diffusion times (Pierpaoli et al. 1996). Neural tissue comprises a specific group of parallel bundles made of axonal bars. The diffusion MRI aims to gain the Three Dimensional (3D) ODF for each voxel (voxel-based analysis of diffusion tensor. i.e., volume of pixel) coordinate. The ODF, which is considered a distorted sphere with a radius of a given direction, is symmetric with the total values of the diffusion of PDF in that direction (Hagmann et al. 2006). The color coding used in ODF according to a diffusion direction for easy visualization, where the red, green, and blue refer to the x, y, and z coordinates, respectively (Topgaard 2017; Shehab et al. 2019a).

2.3.1 Diffusion Contrast

For enabling the measurement of diffusion with MRI, this is done by adding a pair of sharp magnetic field gradient pulses to a conventional MRI sequence (Stejskal and Tanner 1965), such as a simple Pulsed Gradient Spin Echo (PGSE), as shown in Fig. 2.3. The pulse spacing and duration by Δ and δ, respectively. And The gradient vector is denoted by g. The first labels spins of molecules carried through according to their spatial location because the magnetic field differs along one direction. The resulting phase shift is given by

$$\phi_1 = \gamma g. \int_0^{\delta} x(t)dt \tag{2.1}$$

where $x(t)$ is the spin position at time t and γ is the gyromagnetic ratio. The phase shift ϕ_1 is then reverted by the π pulse, thus transformed into $-\phi_1$. Next, the second pulse produces a phase shift ϕ_2:

$$\phi_1 = \gamma g. \int_{\Delta}^{\Delta+\delta} x(t)dt \tag{2.2}$$

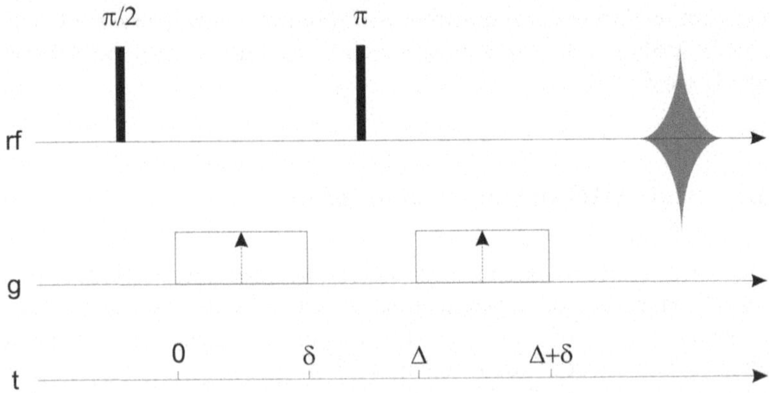

Fig. 2.3 Time diagram of the Stejskal and Tanner pulsed gradient spin echo (PGSE) sequence. Δ and δ are the pulse spacing and pulse duration, respectively (Heisel et al. 2012)

resulting in a net dephasing $\delta(\phi)$:

$$\delta(\phi) = \phi_2 - \phi_1 \tag{2.3}$$

For a moment, let us consider that the pulse duration δ is negligible compared to the pulse spacing, that is $\delta << \Delta$, also called narrow pulse approximation. In this case, we can show that the net dephasing is proportional to the spin displacement r, as follows:

$$\delta(\phi) = \gamma\delta g.[x(\Delta) - x(0)] = q.r \tag{2.4}$$

where $q = \gamma\delta g$ is defined as the gradient wave vector. Therefore, b-value can introduce as:

$$b = \gamma^2 g^2 \delta^2 (\Delta - \delta/3) \tag{2.5}$$

2.3.2 *Diffusion Tensor Analysis*

Mathematically, the tensors are functions or simply arrays of numbers, which depend on specific rules to transform and change coordinates (Bertschinger 2000; Shehab et al. 2017b). Physically, tensors describe the characteristics of a physical system (Stratton 2007). Medically, tensors are the simplest way to represent each voxel in the brain (Wedeen et al. 2005).

The basic premise of diffusion MRI is the differentiation between tissues in the brain based on how water molecules diffuse through each tissue, the motion of diffusion is random. This movement differs among locations, depending on the type of selected tissue. For example, in the gray matter, the diffusion is unrestricted.

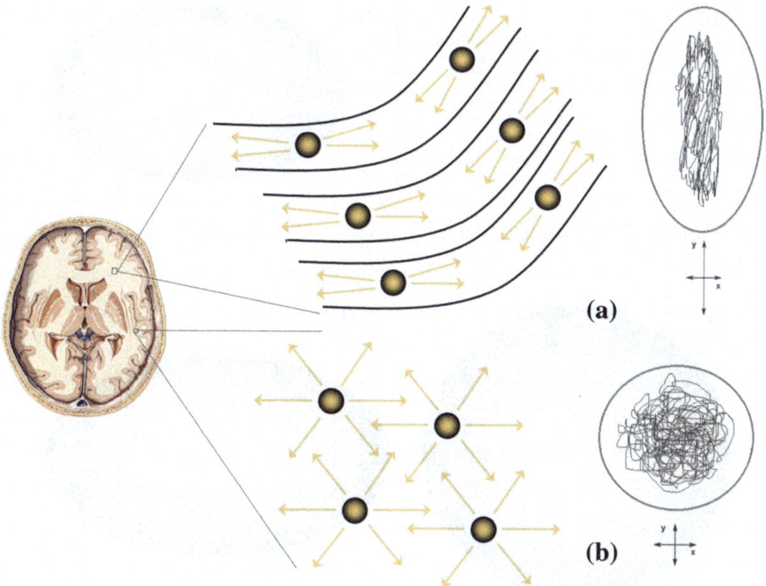

Fig. 2.4 **a** Restricted diffusion (anisotropic). **b** Unrestricted diffusion (isotropic) (Drobyshevsky et al. 2007)

Hence, the water molecules move in any possible direction; this movement is called an isotropic diffusion. The diffusion in the white matter is restricted because of the cover that protects the axons, which allows the water molecules to move freely parallel to the axons (Kleinnijenhuis 2014). This movement is called anisotropic diffusion because of the preferential direction of diffusion, as shown in Fig. 2.4. Thus, we can differentiate between gray and white matters based on whether the diffusion at one voxel in the brain is isotropic or anisotropic. Within different voxels in the white matter, the differential diffusion direction can also determine the orientation of the fiber bundle that goes through any specific voxel.

2.3.2.1 Representation Diffusion Tensor

Large amount of information is contained in the brain image. This section presents this information by using tensors, which represent each voxel in the brain. In each voxel, the following will be determined:

- Is this white matter?
- If yes, what pathway(s) is it part of?
  - What is the orientation of diffusion?
  - What is the magnitude of diffusion?

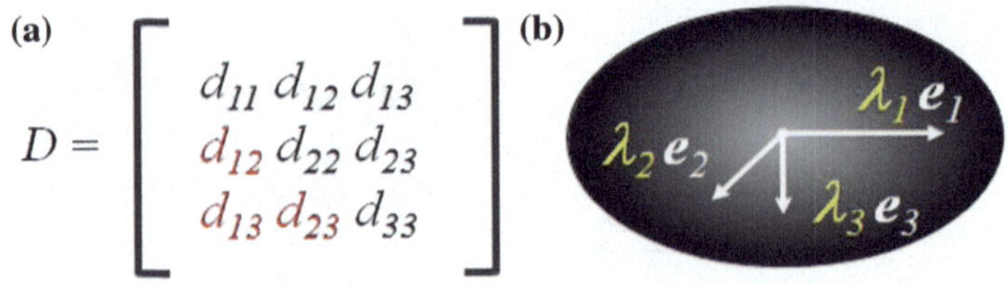

Fig. 2.5 **a** Tensor D: 3 × 3 symmetric matrix, **b** Tensor D with its Eigenvectors and Eigenvalues (Jellison et al. 2004a)

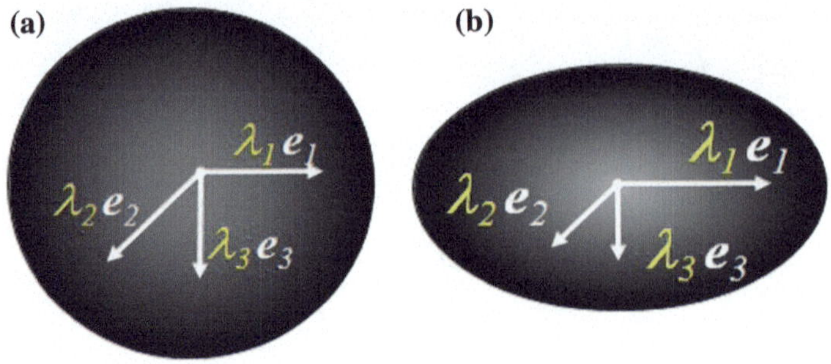

Fig. 2.6 **a** Isotropic. **b** Anisotropic

Simply, a tensor is a 3×3 symmetric matrix. Symmetric means that the upper triangle part is identical to the lower triangle part. Therefore, a symmetric matrix has six distinct elements, as shown in Fig. 2.5a. These elements should be estimated at each voxel in the brain to fit a tensor in each voxel. The matrix should be decomposed into three positive eigenvalues $\lambda_1, \lambda_2, \lambda_3 > 0$ and orthogonal eigenvectors e_1, e_2, e_3 to visualize it, as shown in Fig. 2.5b.

Eigenvectors determine the coordinate system, and eigenvalues are the links of each of these vectors. For each voxel in the brain, a tensor is estimated based on a DWI, and the corresponding tensors can be grouped into these three eigenvectors and three eigenvalues. The physical interpretation used to determine each voxel belongs to either isotropic $\lambda_1 \approx \lambda_2 \approx \lambda_3$ (Fig. 2.6a) or anisotropic diffusion $\lambda_1 >> \lambda_2 \approx \lambda_3$ (Fig. 2.6b) (Bastin et al. 1998). Hence, the isotropic diffusion moves randomly, and the anisotropic one moves depending on the magnitude of the largest value between three eigenvalues. Eigenvectors express the diffusion direction, and eigenvalues express the diffusion magnitude.

The determination of voxel in the white or gray matter depends on some mathematical equations. The Mean Deviation (MD) of the three eigenvalues for the required voxel is calculated (Eq. 2.6), and the obtained value is used to calculate the Fractional Anisotropy (FA), as shown in Eq. 2.7.

$$MD(i) = [\lambda_1(i) + \lambda_2(i) + \lambda_3(i)]/3 \tag{2.6}$$

$$FA(i)^2 = \frac{3}{2}\frac{[\lambda_1(i) - MD(i)]^2 + [\lambda_2(i) - MD(i)]^2 + [\lambda_3(i) - MD(i)]^2}{\lambda_1(i)^2 + \lambda_2(i)^2 + \lambda_3(i)^2} \tag{2.7}$$

where i is a required voxel.

The variance of the three eigenvalues for FA ranges within (0, 1), thereby $0 \leq FA \leqslant 1$ (Beppu et al. 2003; Shehab et al. 2017c). Thus, when the value of FA is close to 0, the three eigenvalues are similar to one another, this refer to gray matter area (isotropic). By contrast, when the value close to 1, the eigenvalues are different from one another, this refer to white matter area (anisotropic).

2.4 Fundamental MRI Techniques

2.4.1 Diffusion Weighted Imaging

DWI is an imaging technique that can probe tissue microstructure in the living human brain. DWI can detect varying levels of anisotropic diffusion that could alter white matter integrity in diseased tissue. Therefore, this technique is a distinct method to characterize between fiber directions in the white matter area. Figure 2.7 shows a DWI, in which the dark area refers to high diffusion along the specific path, whereas the light area refers to less diffusion (Le Bihan 2003). The method is not as informative as having an image with the ODF of the diffusion in each voxel, but DWI is still applied clinically for stroke investigations, thereby causing damage to neural pathways and restrictions in water movement.

2.4.2 Diffusion Tensor Imaging

DTI, also known as DT-MRI, is attracting increasing interest after introduced by Basser et al. (1994) the diffusion tensor. Such interest is caused by the ability of this technique to quantify the magnitude of diffusion. In addition, it measures the direction of the tissue for anisotropy independently. DTI has b-value roughly between 700 and 1300 s/mm^2, where the common value b-value equal 1000 s/mm^2 with 6 gradient directions (Alexander et al. 2007).

In DTI, the primary eigenvector of the tensor can be visualized using two methods. The first way FA map, this method is picking out of the eigenvector that corresponds to the highest eigenvalue. Consequently, the preferential direction of water molecules diffusion is obtained. The eigenvector at each voxel in the brain is shown in Fig. 2.8a. The second way is the use of RGB map. In this method, the motion of tensor related to RGB map is observed. For example, when the voxel's color is red, the preferential

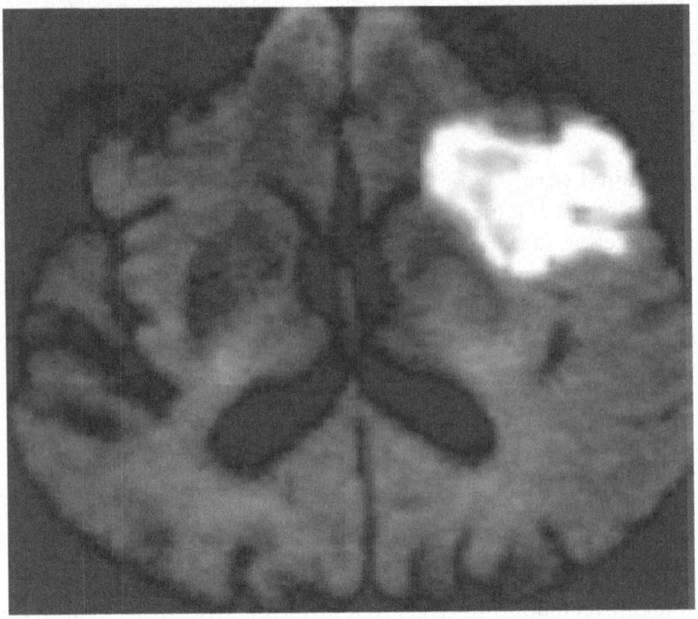

Fig. 2.7 Diffusion Weighted Imaging (DWI) (Fat et al. 2013)

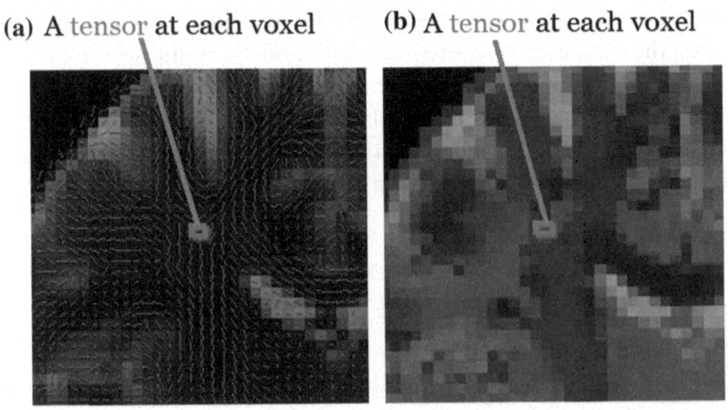

Fig. 2.8 **a** FA map, the direction of eigenvector corresponding to greatest eigenvalue, **b** RGB map, color indicates direction as follows: red, left-right; green, anteroposterior; blue, superior-inferior. This convention applies to all the direction (Rosenbloom et al. 2003)

direction of diffusion specific voxel is mostly left to right. When the voxel's color is green, the preferential direction of diffusion of a specific voxel is mostly antero-posterior. When the voxel's color is blue, the preferential direction of diffusion of a specific voxel is mostly superior to inferior (Jellison et al. 2004b), as shown in Fig. 2.8b.

2.4.3 *High Angular Resolution Diffusion Imaging*

Multiple fiber populations crossing within the same voxel cannot be addressed by DTI technique because of the six directions at lower b-values (low b-value leads to low accuracy). Therefore, imaging techniques require high angular resolution. High Angular Resolution Diffusion Imaging (HARDI) scheme deals with more than 50 gradient directions at high b-value (Berman et al. 2013), where b-value measures the degree of diffusion weight applied. To determine the slow-moving water molecules and smaller diffusion distances, b-value should be higher (e.g., in HARDI, b-value > 4000 sec/mm^2). Thus, the number of acquisitions is increased to obtain further diffusion information. Accordingly, the measurement accuracy is increased, and the direction with high accuracy is established for the selected axon.

In HARDI, the ODF for describing the diffusion profile can have multiple maxima and thus capture complex fiber structures, such as fanning, crossing, kissing, and bending fibers (Fig. 2.9). ODF can deal with multiple fiber orientations within a voxel. The ODF can be calculated in two ways: fiber ODF (with MRI signal profile with single fiber response function in signal space) and diffusion ODF (with q-space technique with single fiber response function in diffusion ODF space) (Anderson 2005; Tournier et al. 2007; Descoteaux et al. 2009a). Both of these techniques have uncertainties associated with estimated fiber orientations. Hence, most researchers used probabilistic methods when selecting for the orientation. The following HARDI techniques (Campbell et al. 2014) applied the ODF in tractography.

2.4.3.1 Diffusion Spectrum Imaging

DSI is used to reconstruct fiber pathways of the brain, hence providing a graphical picture of the "connectome". DSI can resolve multiple intravoxel fiber directions (Wedeen et al. 2005). It is a model-free approach that allows to measure diffusion without making any assumption about the form of the underlying diffusion function. However, a controversy exists in the literature regarding the accuracy of the diffusion tractography reconstruction (Connerney 2014). DSI has the highest b-value comparing with the previous techniques, which is more than 8000 sec/mm^2. The main drawbacks to routine clinical use of DSI still remain in long acquisition times and the need for strong magnetic field gradients.

Fig. 2.9 Fibre Architectures (Sotiropoulos and Zalesky 2017)

2.4.3.2 Q-Ball Imaging

QBI endeavors to catch a rough photo of the diffusion PDF in each voxel without acquiring the numerous disadvantages that underlying q-space imaging techniques introduced (e.g., long scan times) (QBI) (Tuch 2004b). As mentioned previously, the diffusion PDF being identified with the diffusion signal by a Fourier relationship, an immediate reconstruction of that diffusion rule can be gotten by a Fourier change of a three dimensional sampling of q-space (Nichols and Holmes 2004). The three dimensional PDF provides a complete picture of the tissue microstructure however for motivations behind introduction structure mapping of the tissue, an estimation of that PDF showing only the introduction structure characteristic in the diffusion function would be adequate. This is portrayed through the diffusion ODF. Getting the ODF from a three dimensional grid examining of q-space has various restrictions however. ODF extraction requires express estimation of the radial projections which would present artifacts in the mapping from Cartesian to spherical directions Abualigah and Hanandeh (2015). It will be considerably more efficient to gauge the diffusion ODF by sampling the diffusion signal on a spherical harmonic in diffusion proportional space, directly. This structures the premise of QBI. Of the primary attributes that lead to quick interest of the QBI technique are model independence and linear estimation.

QBI, is a generally utilized remaking plan for HARDI, from which ODFs are approximated by the reversal spherical tomographic called the Funk-Radon transform (FRT) (Funk 1915). This present procedure's straightforwardness and its capacity to determine intravoxel fiber introductions have made it mainstream for fiber following and portraying white matter architecture. Various as of late proposed strategies have transformed QBI into an extremely productive and vigorous system (Anderson 2005; Tristán-Vega et al. 2009). The following points include more details about the ODF and spherical harmonic

- **Orientation Distribution Function**
  The orientation distribution function (ODF) is a 3D spherical function founded to detecte the dominant fiber orientations in the underlying voxel. In general, the

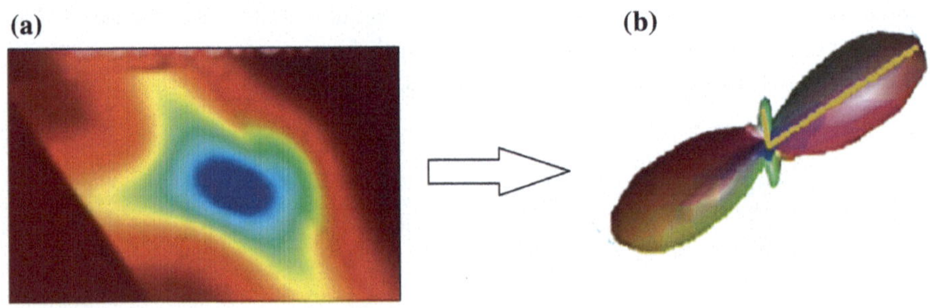

Fig. 2.10 **a** PDF: Can reval the distribution of fiber directions and information about "fast" and "slow" diffusion. **b** ODF: Estimated by the radial integral of displacement PDF (Hagmann et al. 2006)

Fig. 2.11 The effect of the magnitude of regularization parameter from left to right; under-regularized, optimally regularized, and over regularized cases (Juneja 2012)

ODF is the radial projection of the 3D PDF onto the unit sphere. Wedeen et al. (2000) first coined the term ODF. It was presented in order to by pass the need to estimate the whole PDF which requires excessive scan times and extensive computing requirements, Fig. 2.10 illustrated how to move from PDF to ODF. The ellipsoid formed by the eigenvalues and eigenvectors of the diffusion tensor is considered as one of the most basic ODFs. The ODF only describes the areas where the PDF has the most cluster by forming peaks in those directions. These peaks are supposed to correspond with the underlying fiber directions (Alexander 2005b; Shehab et al. 2019b). The peaks comfortable with the underlying fiber directions is of importance, where DTI considered as a standard method that breaks down in areas of the brain where there is partial voluming in a given voxel. As mentioned above, DTI's underlying assumption is that the underlying diffusion occurring is Gaussian in nature and the tensor has only one single peak so providing just a single fiber orientation estimate in every voxel. In the case of multiple fiber crossings within a voxel, the diffusion PDF will show multiple peaks which the tensor in DTI will fail to capture. Therefore, the ODF can deal with this case to complete representation of the complexity in some regions in the white matter Abualigah et al. (2016b).

The regularization parameter controls the smoothness of the generated fiber ODF, and the optimal value may vary from case to case. It provides the strongest robustness to noise and the most accurate detection of fiber crossings (Deriche et al. 2009). It may require several trial-and-error to figure out the best setting. Figure 2.11 shows the effects of the magnitude of regularization parameter on the ODF. It is worth mentioning that the value of the regularization parameter lies between 0 and 1, and the popular value around 0.006 (Descoteaux et al. 2007).

- **Q-Ball Imaging Using the Spherical Harmonic Basis**

  The single spherical shell in diffusion reciprocal space is usually used to acquire HARDI data. Hess et al. (2006a) benefited from the spherical geometry in the data to reconstruct the ODF through introducing an implementation of QBI using the spherical harmonic. Based on that, the ODF is represented as a linear integration of spherical harmonics. It is worth mentioning, that there is a relation between a harmonic model order number and number of measured data points. Where using the larger series orders will obtain the highest possible angular resolution. However, a higher frequency harmonics will be less reliably estimated from determined data Abualigah et al. (2016a). So, the emergence of spurious peaks in the ODFs reconstructed (Roine et al. 2015). In several researchers attempts to reduction the sensitivity to noise. For example, when a large harmonic order approximation was desired, then they applied matrix regularization based on the Tikhonov method (Hess et al. 2005). However, the selection of the regularization parameter value was left at the discretion the investigator.

Thereafter, Descoteaux et al. (2007) used a regularized analytical solution with the Laplace-Beltrami operator for QBI in the spherical harmonic basis. This operator is known on the unit sphere and is a criterion of smoothness for functions on the unit sphere. The authors proved that the performance of the Laplace-Beltrami regularization is better than the Tikhonov regularization (Shehab et al. 2018b). In other words, in the experimental simulations, the authors used set a priori in the smoothing parameter value to provide a good separation of single fiber from two fibers distributions over a large range of SNR and b-values (Descoteaux et al. 2006).

2.5 Previous Related Work

In diffusion MRI, the ODF is used to determine the probability of the diffusion in specific directions. Through obtaining the ODF of the water diffusion for each voxel coordinate. Many methods have been introduced in the literature to represent the measured signals and to resolve multiple fiber orientations within voxels. This section overviews the literatures related to the ODF.

2.5.1 Maximum Entropy

The Fourier transform of a three-dimensional function and maximum entropy parametrization are used to define a statistic called the persistent angular structure (PAS) (Jansons and Alexander 2003; Abualigah et al. 2017b). Where, the samples of the Fourier transform responsible for identifying the probability density function of particle displacements. Then, the representation of the relative mobility of particles in each direction done by PAS which is calculated in each voxel. However, the numerical approximations are very critical for the local minima. Thus, it's not guaranteed for covering the global minima. Peled et al. (2006) determined the coordination of the frame for single tensor, this method achieves good estimation with two tensors residing. But, it's not accurate to estimate more than two tensors. While, Bergmann et al. (2006) took advantage of the weight for each signal with the distance from the equator of the tensors to estimate the ODF profile.

Maréchal (2015) used maximum entropy principle to measure the discrepancy between the probability to be inferred and some reference measure. The solution is computed via the unconstrained maximization of a smooth concave function, whose number of variable is merely (twice) the number of Fourier samples. While, Cheng et al. (2014) employed the maximum entropy with spherical deconvolution and Cartesian tensor ensure non-negativity throughout the unit sphere. Nonetheless computationally intensive and are susceptible to errors caused by numerical spherical integration.

2.5.2 Radial Basis Functions

Radial Basis Function (RBF) is used by Tuch et al. (2003) and Tuch (2004b) to measure the signal and obtain intensive sampling on the sphere based on the Funk-Radon transform. This method provides the efficient and accurate for measuring the signal of the ODF. However, it needs lots of coefficients to represent the ODF. Also, no natural way to define the principal diffusion directions from the coefficients themselves. Therefore, it requires using other methods to defined the maxima of the ODF. As well, to model the measured signal, Kaden et al. (2007) proposed the mixture of Bingham with Bayesian approach to determine the parameter of this model.

Representing the diffusion signal as a linear combination of directional RBF scattered in q-space for estimating the diffusion propagator by Ning et al. (2015). In particular, authors focus on a special case of anisotropic Gaussian basis functions and derive analytical expressions for the ODF. In-vivo human brain data and a physical phantom with known crossing angle are used to validate proposed method.

2.5.3 Spherical Harmonics

Spherical harmonics utilized by many researchers such as Chen et al. (2004), Descoteaux et al. (2007), Parker et al. (2013) and Yeh and Tseng (2013) for estimating the ODF. This method starts with computing the coefficients of the spherical harmonic and determine the best fit measured. Then, obtained the desired ODF through subsequent modification of the coefficient. While Descoteaux et al. (2007) added smoothness constraint by proposed a regularized version to make it strong to noise. However, in case choose the low-order which mean more smooth. So, it may loose of direction information. Also, in the case of choose high-order, it will need many coefficients to represent the ODF. Knowing that, with the coefficients there is no direct connection between the principles of the diffusion. And it important to know the maxima to obtain the diffusion direction, which will cause an error in case don't know the maxima.

2.5.4 Other Approaches Reported in the Literature

The generalized diffusion tensor formalism is able to solve the underlying spin displacement for the complex area with high accuracy. So, Liu et al. (2004), Shehab et al. (2018a) utilized this method with the non-Gaussian diffusion to represent the fiber clearly in the complex areas. While Parker and Alexander (2003) used the mixture of Gaussian model that represent multi-fiber direction at each voxel. For the same purpose, the spherical deconvolution is used with non-negative constraint to ensure there are no negative values for direction (Tournier et al. 2008; Qi et al. 2013; Cheng et al. 2014). Whereas Kaden et al. (2008) and Alexander (2005a) applied the spherical deconvolution to determine the fiber orientation density. The detecting of the multiple diffusion directions done through estimating of diffusion orientation transform, the mixture of Wishart distribution and high-order-tensor by Tan et al. (2015), Jian and Vemuri (2007) and Ghosh et al. (2008a) respectively.

Another related work is by McGraw et al. (2006), where the authors propose ways to interpolate and compute distances between ODF's. They, however, do not address the problem of computing the ODF from the measured signals, but assume the ODF's to be precomputed using any known method. The known ODF is then represented as a mixture of von Mises-Fisher distributions, with the mixture parameters computed using an EM-style algorithm (Banerjee et al. 2005; Shehab et al. 2017a). Also, Peled et al. (2006) used the coordinate frame of a single tensor fit to estimate two tensors residing in the resulting plane. It is not however known if the method can be extended to estimate more than two tensors. Furthermore, Bergmann et al. (2006) provides a way to estimate the ODF profile by weighting each signal inversely with the distance from the equator. They further use a binary integer programming approach to estimate the different tensors from the computed ODF. While the authors have shown nice results for 321 gradient directions, one would have to interpolate the measured signal if the sampling of the signal is sparse.

Savadjiev et al. (2006) were among the first to propose a technique that applies to the regularization of HARDI data. The authors introduce the concept of co-helicity, which they use to obtain parameters of curves passing through a given point. Their goal is to improve the directional support within diffusion profiles in terms of local neighborhood. The authors validate their results against ground truth from synthetic data and HARDI data acquired from a biological phantom.

2.6 Summary

This chapter presents an overview of the MRI and the principle of the diffusion MRI work have been mentioned in detail. In particular, the diffusion MRI in the human brain through covering how to measure the diffusion MRI and the diffusion tensor analysis. Characterization of the MRI techniques with their advantages and disadvantages (i.e., DWI, DTI, DSI, and QBI) were reviewed with illustrated each of ODF and spherical harmonic. Finally, the previous related works were described.

References

Abualigah, L. M. Q., & Hanandeh, E. S. (2015). Applying genetic algorithms to information retrieval using vector space model. *International Journal of Computer Science, Engineering and Applications, 5*(1), 19.

Abualigah, L. M., Khader, A. T., & Al-Betar, M. A. (2016a). Multi-objectives-based text clustering technique using k-mean algorithm. In *2016 7th International Conference on Computer Science and Information Technology (CSIT)*, pp. 1–6. IEEE.

Abualigah, L. M., Khader, A. T., & Al-Betar, M. A. (2016b). Unsupervised feature selection technique based on genetic algorithm for improving the text clustering. In *Computer Science and Information Technology (CSIT), 2016 7th International Conference on*, pp. 1–6. IEEE.

Abualigah, L. M., Sawaie, A. M., Khader, A. T., Rashaideh, H., Al-Betar, M. A., & Shehab, M. (2017). β-hill climbing technique for the text document clustering. *New Trends in Information Technology, 60.*

Alexander, D. C. (2005a). Maximum entropy spherical deconvolution for diffusion mri. In *Biennial International Conference on Information Processing in Medical Imaging*, pp. 76–87. Springer.

Alexander, D. C. (2005b). Multiple-fiber reconstruction algorithms for diffusion mri. *Annals of the New York Academy of Sciences, 1064*(1), 113–133.

Anderson, A. W. (2005). Measurement of fiber orientation distributions using high angular resolution diffusion imaging. *Magnetic Resonance in Medicine, 54*(5), 1194–1206.

Alexander, A. L., Lee, J. E., Lazar, M., & Field, A. S. (2007). Diffusion tensor imaging of the brain. *Neurotherapeutics, 4*(3), 316–329.

Assaf, Y., & Pasternak, O. (2008). Diffusion tensor imaging (dti)-based white matter mapping in brain research: A review. *Journal of Molecular Neuroscience, 34*(1), 51–61.

Banerjee, A., Dhillon, I. S., Ghosh, J., & Sra, S. (2005). Clustering on the unit hypersphere using von mises-fisher distributions. *Journal of Machine Learning Research, 6*(Sep), 1345–1382.

Barmpoutis, A., & Vemuri, B. C. (2009). Information theoretic methods for diffusion-weighted mri analysis. In *Emerging Trends in Visual Computing*, pp. 327–346. Springer.

Basser, P. J., Mattiello, J., & LeBihan, D. (1994). Mr diffusion tensor spectroscopy and imaging. *Biophysical Journal, 66*(1), 259.

Bastin, M. E., Armitage, P. A., & Marshall, I. (1998). A theoretical study of the effect of experimental noise on the measurement of anisotropy in diffusion imaging. *Magnetic Resonance Imaging, 16*(7), 773–785.

Beppu, T., Inoue, T., Shibata, Y., Kurose, A., Arai, H., Ogasawara, K., et al. (2003). Measurement of fractional anisotropy using diffusion tensor mri in supratentorial astrocytic tumors. *Journal of Neuro-Oncology, 63*(2), 109–116.

Bergmann, Ø., Kindlmann, G., Lundervold, A., & Westin, C.-F. (2006). Diffusion k-tensor estimation from q-ball imaging using discretized principal axes. In *International Conference on Medical Image Computing and Computer-Assisted Intervention*, pp. 268–275. Springer.

Berman, J. I., Lanza, M. R., Blaskey, L., Edgar, J. C., & Roberts, T. P. L. (2013). High angular resolution diffusion imaging probabilistic tractography of the auditory radiation. *American Journal of Neuroradiology, 34*(8), 1573–1578.

Bertschinger, C. E. (2000). Introduction to tensor calculus for general relativity.

Campbell, J. S. W., MomayyezSiahkal, P., Savadjiev, P., Leppert, I. R., Siddiqi, K., & Pike, G. B. (2014). Beyond crossing fibers: bootstrap probabilistic tractography using complex subvoxel fiber geometries. *Frontiers in neurology, 5.*

Chen, Y., Guo, W., Zeng, Q., Yan, X., Huang, F., Zhang, H., He, G., Vemuri, B. C., & Liu, Y. (2004). Estimation, smoothing, and characterization of apparent diffusion coefficient profiles from high angular resolution dwi. In *Computer Vision and Pattern Recognition, 2004. CVPR 2004. Proceedings of the 2004 IEEE Computer Society Conference on*, vol. 1, pp. I–588. IEEE.

Cheng, J., Deriche, R., Jiang, T., Shen, D., & Yap, P.-T. (2014). Non-negative spherical deconvolution (nnsd) for estimation of fiber orientation distribution function in single-/multi-shell diffusion mri. *NeuroImage, 101*, 750–764.

Connerney, M. (2014). *3D reconstruction of motor pathways from tract tracing in the rhesus monkey*. PhD thesis, school of medicine thesis 3D reconstruction of motor pathways from tract tracing in the rhesus monkey by michael CONNERNEY BA, Boston University.

Deriche, R., Calder, J., & Descoteaux, M. (2009). Optimal real-time q-ball imaging using regularized kalman filtering with incremental orientation sets. *Medical Image Analysis*, *13*(4), 564–579.

Descoteaux, M., Angelino, E., Fitzgibbons, S., & Deriche, R. (2006). Apparent diffusion coefficients from high angular resolution diffusion imaging: Estimation and applications. *Magnetic Resonance in Medicine*, *56*(2), 395–410.

Descoteaux, M., Angelino, E., Fitzgibbons, S., & Deriche, R. (2007). Regularized, fast, and robust analytical q-ball imaging. *Magnetic Resonance in Medicine*, *58*(3), 497–510.

Descoteaux, M., Deriche, R., Knösche, T. R., & Anwander, A. (2009a). Deterministic and probabilistic tractography based on complex fibre orientation distributions. *Medical Imaging, IEEE Transactions on*, *28*(2), 269–286.

Drobyshevsky, A., Derrick, M., Wyrwicz, A. M., Ji, X., Englof, I., Ullman, L. M., Zelaya, M. E., Northington, F. J., & Tan, S. (2007). White matter injury correlates with hypertonia in an animal model of cerebral palsy. *Journal of Cerebral Blood Flow & Metabolism*, *27*(2), 270–281.

Einstein, A. (1956). *Investigations on the Theory of the Brownian Movement*. Courier Corporation.

Fat, M. J. L., Al-Hazzaa, M., Bussière, M., dos Santos, M. P., Lesiuk, H., & Lum, C. (2013). Heparin dosing is associated with diffusion weighted imaging lesion load following aneurysm coiling. *Journal of Neurointerventional Surgery*, *5*(4), 366–370.

Funk, P. (1915). Über eine geometrische anwendung der abelschen integralgleichung. *Mathematische Annalen*, *77*(1), 129–135.

Ghosh, A., Descoteaux, M., & Deriche, R. (2008a). Riemannian framework for estimating symmetric positive definite 4th order diffusion tensors. In *International Conference on Medical Image Computing and Computer-Assisted Intervention*, pp. 858–865. Springer.

Hagmann, P., Jonasson, L., Maeder, P., Thiran, J.-P., Van Wedeen, J., & Meuli, R. (2006). Understanding diffusion mr imaging techniques: from scalar diffusion-weighted imaging to diffusion tensor imaging and beyond 1. *Radiographics*, *26*(suppl_1):S205–S223.

Heidi, J.-B., & Behrens, T. E. J. (2013). *Diffusion MRI: from quantitative measurement to in vivo neuroanatomy*. Academic Press.

Heisel, K. A., Goto, J. J., Krishnan, V. V., et al. (2012). Nmr chromatography: molecular diffusion in the presence of pulsed field gradients in analytical chemistry applications.

Hess, C. P., Mukherjee, P., Han, E. T., Xu, D., & Vigneron, D. B. (2005). A spherical harmonic approach to q-ball imaging. In *Proceedings of the 13th Annual Meeting of ISMRM*, p. 389.

Hess, C. P., Mukherjee, P., Han, E. T., Xu, D., & Vigneron, D. B. (2006). Q-ball reconstruction of multimodal fiber orientations using the spherical harmonic basis. *Magnetic Resonance in Medicine*, *56*(1), 104–117.

Jansons, K. M., & Alexander, D. C. (2003). Persistent angular structure: new insights from diffusion magnetic resonance imaging data. *Inverse problems*, *19*(5), 1031.

Jellison, B. J., Field, A. S., Medow, J., Lazar, M., Salamat, M. S., & Alexander, A. L. (2004a). Diffusion tensor imaging of cerebral white matter: a pictorial review of physics, fiber tract anatomy, and tumor imaging patterns. *American Journal of Neuroradiology*, *25*(3), 356–369.

Jellison, B. J., Field, A. S., Medow, J., Lazar, M., Salamat, M. S., & Alexander, A. L. (2004b). Diffusion tensor imaging of cerebral white matter: A pictorial review of physics, fiber tract anatomy, and tumor imaging patterns. *American Journal of Neuroradiology*, *25*(3), 356–369.

Jian, B., & Vemuri, B. C. (2007). A unified computational framework for deconvolution to reconstruct multiple fibers from diffusion weighted mri. *IEEE transactions on medical imaging*, *26*(11), 1464–1471.

Juneja, V. (2012). Novel phantoms and post-processing for diffusion spectrum imaging.

Kaden, E., Anwander, A., & Knösche, T. R. (2008). Variational inference of the fiber orientation density using diffusion mr imaging. *Neuroimage*, *42*(4), 1366–1380.

Kaden, E., Knösche, R.T., & Anwander, A. (2007). Parametric spherical deconvolution: inferring anatomical connectivity using diffusion mr imaging. *NeuroImage*, *37*(2), 474–488.

Kleinnijenhuis, M. (2014). *Imaging fibres in the brain*, volume 159. Michiel Kleinnijenhuis.
Le Bihan, D. (2003). Looking into the functional architecture of the brain with diffusion mri. *Nature Reviews Neuroscience*, *4*(6), 469–480.
Liu, C., Bammer, R., Acar, B., & Moseley, M. E. (2004). Characterizing non-gaussian diffusion by using generalized diffusion tensors. *Magnetic Resonance in Medicine*, *51*(5), 924–937.
McGraw, R., Vemuri, B. C., Yezierski, B., & Mareci, T. (2006). Von mises-fisher mixture model of the diffusion odf. In *Biomedical Imaging: Nano to Macro, 2006. 3rd IEEE International Symposium on*, pp. 65–68. IEEE.
Nichols, T. E., & Holmes, A. P. (2004). Nonparametric permutation tests for functional neuroimaging. *Human Brain Function*, *2*, 887–910.
Ning, L., Westin, C.-F., & Rathi, Y. (2015). Estimating diffusion propagator and its moments using directional radial basis functions. *IEEE Transactions on Medical Imaging*, *34*(10), 2058–2078.
Ogawa, S., Lee, T.-M., Kay, A. R., & Tank, D. W. (1990). Brain magnetic resonance imaging with contrast dependent on blood oxygenation. *Proceedings of the National Academy of Sciences*, *87*(24), 9868–9872.
Parker, G. J. M., & Alexander, D. C. (2003). Probabilistic monte carlo based mapping of cerebral connections utilising whole-brain crossing fibre information. In *Biennial International Conference on Information Processing in Medical Imaging*, pp. 684–695. Springer.
Parker, G. D., Marshall, D., Rosin, P. L., Drage, N., Richmond, S., & Jones, D. K. (2013). A pitfall in the reconstruction of fibre odfs using spherical deconvolution of diffusion mri data. *Neuroimage*, *65*, 433–448.
Peled, S., Friman, O., Jolesz, F., & Westin, C.-F. (2006). Geometrically constrained two-tensor model for crossing tracts in dwi. *Magnetic Resonance Imaging*, *24*(9), 1263–1270.
Pierpaoli, C., Jezzard, P., Basser, P. J., Barnett, A., & Chiro, G. D. (1996). Diffusion tensor mr imaging of the human brain. *Radiology*, *201*(3), 637–648.
Pierre Maréchal. Duality for maximum entropy diffusion mri. In *AIP Conference Proceedings*, volume 1641, pages 205–212. AIP, 2015.
Qi, L., Gaohang, Y., & Yi, X. (2013). Nonnegative diffusion orientation distribution function. *Journal of Mathematical Imaging and Vision*, *45*(2), 103–113.
Roine, T., Jeurissen, B., Perrone, D., Aelterman, J., Philips, W., Leemans, A., et al. (2015). Informed constrained spherical deconvolution (icsd). *Medical Image Analysis*, *24*(1), 269–281.
Rosenbloom, M., Sullivan, E. V., Pfefferbaum, A., et al. (2003). Using magnetic resonance imaging and diffusion tensor imaging to assess brain damage in alcoholics. *Alcohol Research and Health*, *27*(2), 146–152.
Savadjiev, P., Campbell, J. S. W., Pike, G. B., & Siddiqi, K. (2006). 3d curve inference for diffusion mri regularization and fibre tractography. *Medical image analysis*, *10*(5), 799–813.
Shehab, M., Khader, A. T., & Al-Betar, M. A. (2016). New selection schemes for particle swarm optimization. *IEEJ Transactions on Electronics, Information and Systems*, *136*(12), 1706–1711. https://doi.org/10.1541/ieejeiss.136.1706.
Shehab, M., Khader, A. T., & Al-Betar, M. A. (2017a). A survey on applications and variants of the cuckoo search algorithm. *Applied Soft Computing*.
Shehab, M., Khader, A. T., Al-Betar, M. A., & Abualigah, L. M. (2017b) Hybridizing cuckoo search algorithm with hill climbing for numerical optimization problems. In *Information Technology (ICIT), 2017 8th International Conference on*, pp. 36–43. IEEE.
Shehab, M., Khader, A. T., & Laouchedi, M. (2017c). Modified cuckoo search algorithm for solving global optimization problems. In *International Conference of Reliable Information and Communication Technology*, pp. 561–570. Springer.
Shehab, M., Khader, A. T., & Laouchedi, M. (2018a). A hybrid method based on cuckoo search algorithm for global optimization problems. *Journal of ICT*, *17*(3), 469–491.
Shehab, M., Khader, A. T., Laouchedi, M., & Alomari, O. A. (2018b). Hybridizing cuckoo search algorithm with bat algorithm for global numerical optimization. *The Journal of Supercomputing*, pp. 1–28.

Shehab, M., Daoud, M. Sh., AlMimi, H. M., Abualigah, L. M., & Khader, A. T. (2019a). Hybridizing cuckoo search algorithm for extracting the odf maxima in spherical harmonic representation. *International Journal of Bio-Inspired Computation*, (in press).

Shehab, M., Khader, A. T., & Alia., M. A. (2019b). Enhancing cuckoo search algorithm by using reinforcement learning for constrained engineering optimization problems. In *2019 IEEE Jordan International Joint Conference on Electrical Engineering and Information Technology (JEEIT)*, pp. 812–816. IEEE.

Sotiropoulos, S. N., & Zalesky, A. (2017). Building connectomes using diffusion mri: Why, how and but. *NMR in Biomedicine*.

Stejskal, E. O., & Tanner, J. E. (1965). Spin diffusion measurements: spin echoes in the presence of a time-dependent field gradient. *The Journal of Chemical Physics*, *42*(1), 288–292.

Stratton, J. A. (2007). *Electromagnetic theory*. Wiley.

Tan, Ek. T., Marinelli, L., Sperl, J. I., Menzel, M. I., & Hardy, C. J. (2015). Multi-directional anisotropy from diffusion orientation distribution functions. *Journal of Magnetic Resonance Imaging*, *41*(3), 841–850.

Topgaard, D. (2017). Multidimensional diffusion mri. *Journal of Magnetic Resonance*, *275*, 98–113.

Tournier, J.-D., Calamante, F., & Connelly, A. (2007). Robust determination of the fibre orientation distribution in diffusion mri: non-negativity constrained super-resolved spherical deconvolution. *NeuroImage*, *35*(4), 1459–1472.

Tournier, J.-D., Yeh, C.-H., Calamante, F., Cho, K.-H., Connelly, A., & Lin, C.-P. (2008). Resolving crossing fibres using constrained spherical deconvolution: Validation using diffusion-weighted imaging phantom data. *Neuroimage*, *42*(2), 617–625.

Tristán-Vega, A., Westin, C.-F., & Aja-Fernández, S. (2009). Estimation of fiber orientation probability density functions in high angular resolution diffusion imaging. *NeuroImage*, *47*(2), 638–650.

Tuch, D. S. (2004b). Q-ball imaging. *Magnetic Resonance in Medicine*, *52*(6), 1358–1372.

Tuch, D. S., Reese, T. G., Wiegell, M. R., & Van Wedeen, J. (2003). Diffusion mri of complex neural architecture. *Neuron*, *40*(5), 885–895.

Wedeen, V. J., Reese, T. G., Tuch, D. S., Weigel, M. R., Dou, J. G., Weiskoff, R. M., & Chessler, D. (2000). Mapping fiber orientation spectra in cerebral white matter with fourier-transform diffusion mri. In *Proceedings of the 8th Annual Meeting of ISMRM, Denver*, p. 82.

Wedeen, J. Van, Hagmann, P., Tseng, W.-Y. I., Reese, T. G., & Weisskoff, R. M. (2005). Mapping complex tissue architecture with diffusion spectrum magnetic resonance imaging. *Magnetic resonance in medicine*, *54*(6), 1377–1386.

Yeh, F.-C., & Tseng, W.-Y. (2013). Sparse solution of fiber orientation distribution function by diffusion decomposition. *PloS One*, *8*(10), e75747.

Chapter 3
Cuckoo Search Algorithm

3.1 Introduction

Optimization problem exists in many domains, such as engineering, energy, economics, medical, and computer science. It is mainly concerned with finding the optimal values for several decision variables to form a solution to optimization problem. This solution is considered optimal when the decision maker is satisfied with it. An optimization problem is the minimization or maximization of a suitable decision-making algorithm normally adapted to the approximation methods. The principle of decision making entails choosing between several alternatives. The result of this choice is the selection of the best decision from all choices.

Optimization algorithms developed based on nature-inspired ideas deal with selecting the best alternative based on the given objective function. This chapter is limited to the metaheuristic algorithm which it is a general solver template. Where, it can be adapted for various kinds of optimization problems by properly tweaking its operators and configuring its parameters. To elaborate, each optimization algorithm can be categorized into three classes: evolutionary algorithms (EAs), swarm-based algorithms, and trajectory-based algorithms (Shehab et al. 2017a). EAs mimic the evolutionary principle of survival of the fittest. It normally begins with a set of individuals (i.e., a group of solutions) called population. At each generation, the EA algorithms recombine the preferable characteristics of the current population to come up with a new population that will be selected based on the natural selection principle. Examples of EAs include genetic algorithms (GAs) (Holland 1975), genetic programming (GP) (Koza 1994), differential evolution (DE) (Storn and Price 1996), and harmony search algorithm (HSA) (Geem et al. 2001). On the other hand, swarm-based algorithms mimic the behavior of a group of animals when searching for survival. At each iteration, the solutions are normally constructed based on historical information gained by previous generations (Bolaji et al. 2016). Examples of swarm-based algorithms include artificial bee colony (ABC) (Karaboga 2005), particle swarm optimization (PSO) (James and Russell 1995), firefly algorithm (FA)

M. Shehab, *Artificial Intelligence in Diffusion MRI*, Studies in Computational Intelligence 877, https://doi.org/10.1007/978-3-030-36083-2_3

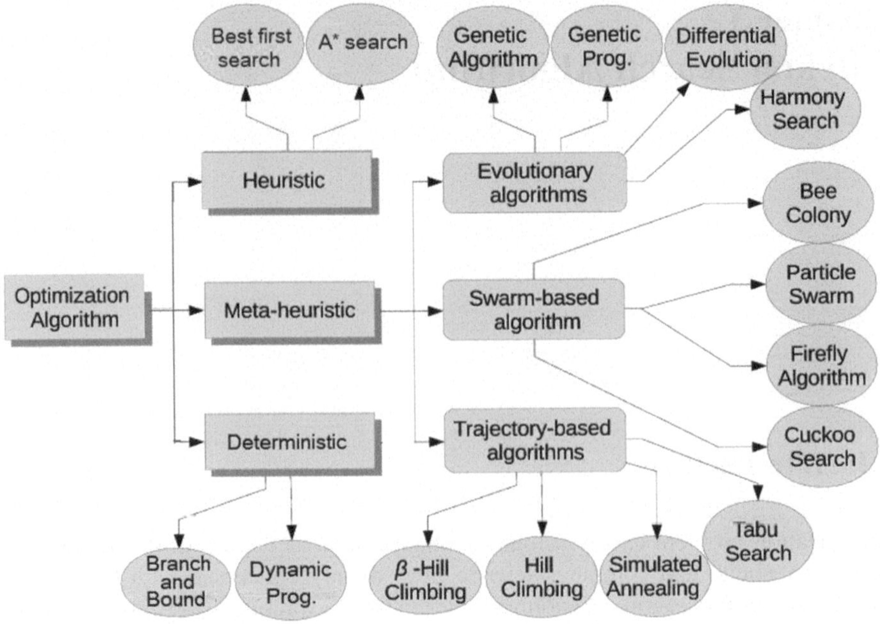

Fig. 3.1 Optimization algorithms

(Yang and Algorithms 2008), and cuckoo search algorithm (CSA) (Yang and Deb 2009). Trajectory-based algorithms start with a single provisional solution. At each iteration, that solution will be moved to its neighboring solution, which resides in the same search space region, using a specific neighborhood structure. Examples of trajectory-based algorithms includes tabu search (TS) (Glover 1977), simulated annealing (SA) (Kirkpatrick et al. 1983), hill climbing (Koziel and Yang 2011), and β-hill climbing (Abualigah et al. 2017) (Fig. 3.1).

The main focus of this chapter, the CSA, was developed by Yang and Deb (2009) to emulate the brood parasitism behavior of cuckoos. Cuckoos have an aggressive breeding behavior, which inspired the optimization algorithm. Brood parasitism is a primary mechanism of cuckoos. This bird lays eggs in the host's nest and carefully matches its eggs through mimicking the pattern and color of the host's eggs (Rajabioun 2011). In case the host recognizes the cuckoo's eggs in its own nest, it will either throw the eggs out or simply leave its nest and build a new one. Therefore, a cuckoo must be accurate in its mimicry of the host eggs, whereas the host must improve its skills in determining the parasitic egg; therein lies the struggle for survival. In the optimization context, each egg in the nest represents a solution, and the cuckoo's egg represents a new solution. The aim is to serve the new and potentially better solutions to replace a not-so-good solution in the nests. The algorithm can be extended to more complicated cases where each nest has multiple eggs representing a set of solutions.

Fraction $P\alpha$ with a probability [0, 1] checks if a host discovers the eggs are not its own (Shehab et al. 2018).

The main merits of the CSA over other optimization algorithms are as follows: no derivation information is required in the initial search, the number of parameters needed to be configured in the initial search is very little, and the inexperienced user can easily interact with it (Shehab et al. 2018a). CSA has a common specification with TAs in exploitation through random walk and with EAs in exploration through $L\acute{e}vy$ flights. It is an efficient metaheuristic algorithm that balances between the local search strategy (exploitation) and the whole space (exploration) (Roy and Chaudhuri 2013), deals with multi-criteria optimization problem, and aims for convergence speed and easy implementation. Owing to these merits, the CSA has been successfully tailored to a wide variety of optimization problems, such as constrained optimization (Ong and Kohshelan 2016), in medical field (Giveki et al. 2012; Liu and Fu 2014; Stewart et al. 2016), clustering and data mining (Goel et al. 2011; Amsaleka and Latha 2014; Cobos et al. 2014), image processing (Pare et al. 2016; Tiwari 2012; Bhandari et al. 2014b, a; Agrawal et al. 2013; Raja and Vishnupriya 2016), economic dispatch problems (Tran et al. 2015; Basu and Chowdhury 2013; Vo et al. 2013; Pham et al. 2016; Sekhar and Mohanty 2016), engineering design (Gandomi et al. 2013a; Ahmed and Salam 2014; Bhargava et al. 2013; Esfandiari 2014; Gandomi et al. 2013b; Shehab et al. 2017c; Kaveh and Bakhshpoori 2016), and power and energy (Ahmed and Salam 2013; Buaklee and Hongesombut 2013; Devabalaji et al. 2016; Elazim and Ali 2016; Femia et al. 2005; Ma et al. 2013; Machowski et al. 2011).

The CSA is also modified and hybridized for the convenience of some combinatorial optimization problems because of the complex nature of some optimization problems (Walton et al. 2011b; Giridhar et al. 2016; Babukartik and Dhavachelvan 2012; Wang et al. 2016b; Lim et al. 2016; Layeb and Boussalia 2012; Shehab et al. 2017b; Noghrehabadi et al. 2011). The parameter setting of the CSA is also addressed by several researchers Tuba et al. (2011), Abdul Rani et al. (2012), Tawfik et al. (2013), Li and Yin (2016).

This chapter provides a comprehensive and exhaustive overview of the theoretical aspects of CSA and presents the readers with sufficient materials for the previous adaptation, modification, and hybridization of the CSA. It also focuses on the principles of CSA, its developments and variants to the original CSA, and a detailed report of recent applications and associated developments attained during the last few years.

This chapter is organized as follows. Section 3.2 introduces the CSA by highlighting its framework. Section 3.3 discusses the procedure of the basic CSA, then the growth of CSA illustrated in Sect. 3.4. After that, the variants of CSA are shown in details in Sect. 3.5, followed by related works of hybridization BCSA in Sect. 7.2. Finally,The conclusion is presented in Sect. 3.7.

3.2 Cuckoo Search Algorithm Inspired from Nature

There are more than 1,000 species of birds in nature, and these birds share some approaches with one another (Rajabioun 2011). For example, all mother birds lay eggs that have different shapes of eggs from one another. Moreover, different nests are built by many birds in isolated places to increase safety from predators (Davies 1970).

Birds that resort to cunning approaches for reproduction, specifically in building nests, are called "brood parasites". These kinds of birds do not build their own nests but rather lay their eggs in the nest of another species, leaving the host to care for its young. The most famous of the brood parasites is the cuckoo. It has a fantastic way in the art of deception. Its strategy involves permeation by removing one egg laid by the host and laying her own. It then carefully matches its egg through mimicking the pattern and color of the host's eggs, a skill that requires high accuracy to ensure its success. The timing of egg-laying is also an amazing way of selecting the nest where the host bird just laid its own eggs (Khan and Sahai 2013). This process will reap benefits after a while; the cuckoo egg will hatch before the host eggs, and the first instinctive action of the host will be to evict its own eggs out of the nest by blind propelling, thus increasing the care and food provided for the cuckoo's chicks. Cunningness is inherited by the chicks; this trait is shown when the chicks mimic the call of host chicks to gain access to more feeding opportunity (Yang and Press 2010).

On the other hand, in case the host recognizes the cuckoo's egg in its nest, they either throw out the strange egg or simply leave their own nest and build a new one. The cuckoo must therefore be more accurate in mimicking the host eggs, whereas the host must improve its skills in determining the parasitic egg. Therein lies the so-called struggle for survival (Shehab et al. 2016).

The use of CSA in the optimization context was proposed by Yang and Deb in 2009. To date, work on this algorithm has significantly increased, and the CSA has succeeded in having its rightful place among other optimization methodologies (Fister Jr et al. 2014; Yang and Deb 2009; Shehab et al. 2019b). This algorithm is based on the obligate brood parasitic behavior found in some cuckoo species, in combination with the *Lévy* flight, which it is a type of random walk which has a power law step length distribution with a heavy tail. It is inspired from behavior discovered of some birds and fruit flies. Also, it has been found (Brown et al. 2007; Pavlyukevich 2007; Shehab and Khader 2018) that *Lévy* flights is an oft-observed random walk in real life (Viswanathan et al. 1999, 2002). The CSA is an efficient metaheuristic swarm-based algorithm that efficiently strikes a balance between local nearby exploitation and global-wide exploration in the search space problem (Roy and Chaudhuri 2013).

The cuckoo has a specific way of laying its eggs to distinguish it from the rest of the birds (Yang and Deb 2014). The following three idealized rules clarify and describe the standard cuckoo search:

- Each cuckoo lays one egg at a time and dumps it in a randomly chosen nest.
- The best nests with high-quality eggs will be carried over to the next generations.
- The number of available host nests is fixed, and the egg laid by a cuckoo is discovered by the host bird with a probability $P\alpha \in (0, 1)$. In this case, the host bird can either get rid of the egg or simply abandon the nest and build a completely new nest. In addition, probability $P\alpha$ can be used by the n host nest to replace the new nests.

3.3 The Procedure of Basic Cuckoo Search Algorithm

The basic CSA procedure is established by Yang and Deb (2009), the founders of CSA. Figure 3.2 shows a flowchart of the CSA. Similar to other swarm-based algorithms, the CSA starts with an initial population of n host nests. These initial host nests will be randomly attracted by the cuckoos with eggs and also by random *Lévy* flights to lay the eggs. Thereafter, nest quality will be evaluated and compared with another random host nest. In case the host nest is better, it will replace the old host nests. This new solution has the egg laid by a cuckoo. If the host bird discovers the egg with a probability $P\alpha \in (0, 1)$, the host either throws out the egg, or abandons it and builds a new nest. This step is done by replacing the abundant solutions with the new random solutions.

Yang and Deb used a certain and simple representation for the implementation, with each ***egg*** representing a solution. As the ***cuckoo*** lays only one egg, it also represents one solution. The purpose is to increase the diversity of new, and probably better, cuckoos (new solutions) and replace them instead with the worst solutions. By contrast, the CSA can be more complicated by using multiple eggs in each nest to represent a set of solutions.

The CSA, as a bat algorithm (Yang 2010b) and an FA (Yang 2010a), uses a balance between exploration and exploitation. The CSA is equiponderant to the integration of a *Lévy* flights. When generating new solutions x^{t+1} for, say, a cuckoo i, a *Lévy* flight is performed

$$x_i^{t+1} = x_i^t + \alpha \bigoplus Le'vy(\lambda) \tag{3.1}$$

where $\alpha > 0$ is the step size which should be related to the scales of the problem of interests. In most cases, we can use $\alpha = 1$. The x_i^t in the Eq. 3.1 represents the current location, which is the only way to determine the next location x_i^{t+1}. This is called the random walk and the Markov chain. The product $\bigoplus$ means entrywise multiplications. This entrywise product is similar to those used in PSO, but here the random walk via *Lévy* flight is more efficient in exploring the search space as its step length is much longer in the long run. A global explorative random walk by using *Lévy* flights can be expressed as follows:

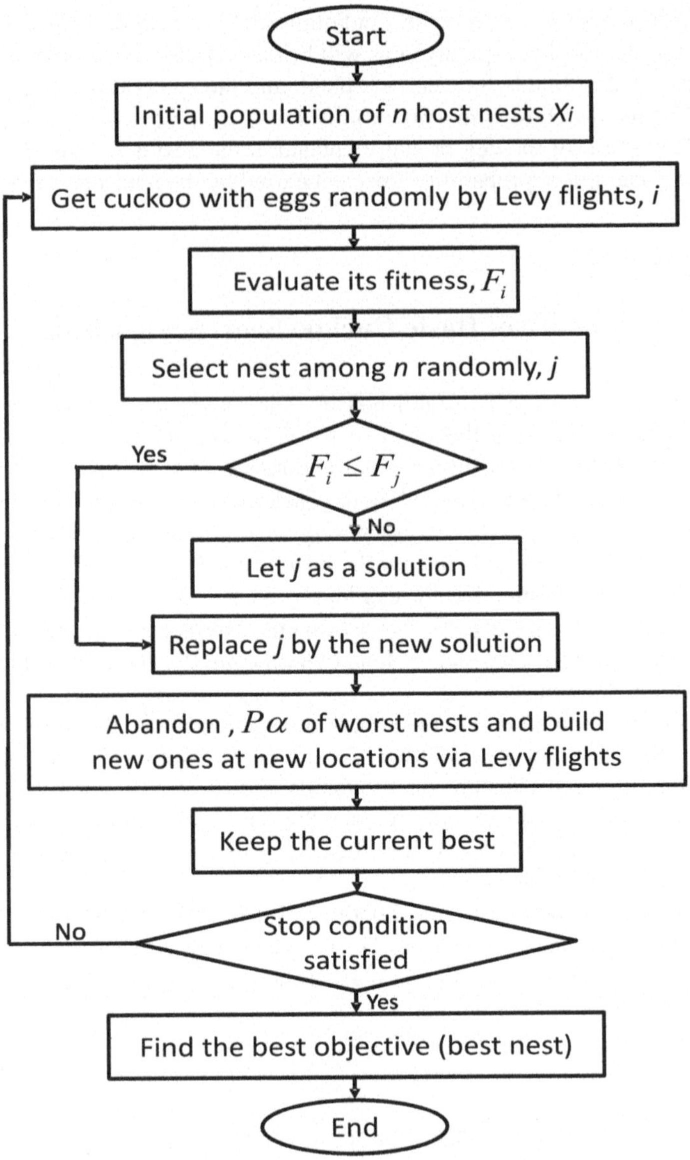

Fig. 3.2 Flowchart of Cuckoo Search Algorithm (Abdul Rani et al. 2012)

$$L\grave{e}vy \sim u = t^{-\lambda}, \quad 1 < \lambda \leqslant 3 \tag{3.2}$$

where λ is a parameter which is the mean or expectation of the occurrence of the event during a unit interval. Here the steps essentially form a random walk process with a power law step-length distribution with a heavy tail. Some of the new solutions should be generated by $L\acute{e}vy$ walk around the best solution obtained so far, this will speed up the local search. However, a substantial fraction of the new solutions should be generated by far field randomization and whose locations should be far enough from the current best solution, this will make sure the system will not be trapped in a local optimum (Shehab et al. 2019a). Algorithm 1 shows the representation of the CSA search process.

Algorithm 1 Basic Cuckoo Search algorithm

Objective function $f(X), X = (x_1, ...x_d)^T$
Generate initial population of n host nests $Xi(i = 1, 2, ..., n)$
while $t < Max\_iterations$ **do**
    $(t < MaxGeneration)$ or (stop criterion)
    Get a cuckoo randomly by Levy flights
    evaluate its quality $fitness F_i$
    Choose α nest among n (say, j) randomly
    if $F_i > F_j$ **then**
        replace j by the new solution;
    end if
    A fraction $(P\alpha)$ of worse nests are abandoned and new ones are built;
    Keep the best solutions (or nests with quality solutions);
    Rank the solutions and find the current best
end while
Postprocess results and visualization

3.4 The Growth of Cuckoo Search Algorithm in the Literature

Several papers have been widely published on the cuckoo search algorithm. In this section, the materials have been collected based on using the CSA as a keyword through two stages. Firstly, the published CSA papers are obtained from the highly-reputed publishers such as Elsevier, Springer, and IEEE as well, from other journals which were searched using Google Scholar. Second, the search results are classified based on the year of publication as shown in this section, and the subject as shown in the application section and theoretical section.

In 2009, CSA was developed and evaluated using multimodal objective functions and then compared with GA and PSO (Yang and Deb 2009, 2010). In 2011, it gradually started to develop, with the enhancement of the convergence rate and the accuracy of the standard CSA utilized for training the feedforward neural networks (Valian et al. 2011). Subsequently, in 2011, Tuba et al. (2011), Walton et al. (2011b) implemented a modified CSA by determining the step size from the sorted rather than only the permuted fitness matrix and by adding information exchange between the best solutions, respectively.

In 2012, CSA began to be used in many problems. Durgun and Yildiz (2012) proposed CSA for solving structural vehicle design optimization problems, which improves fuel efficiency and cost. In addition, Xiang-Tao and Ming-Hao (2012) applied CSA in combination with orthogonal learning to estimate the parameters of chaotic systems. In the same year, the same authors applied the binary version of CSA. The same binary versions are used to solve 0–1 knapsack problems (Gherboudj et al. 2012). For evaluation purposes, the performance of the CSA binary version is superior compared with the binary PSO.

In 2013, CSA was also used for the optimization design of truss structure problems (Gandomi et al. 2013a) and the optimization of machining parameters in milling operations (Yildiz 2013). Both works obtained very effective and robust outcomes compared with GA, ant colony optimization (ACO), and PSO. The problem of optimum synthesis of a six-bar double-dwell linkage was successfully tackled using CSA in mechanical engineering (Bulatović et al. 2013). CSA has likewise been successfully tailored for structural engineering optimization problems, where the obtained solutions are better than the best solutions obtained by the existing methods by a large gap (Talatahari et al. 2013).

Major studies on CSA were conducted in 2014 across different disciplines, where different hybridizations and modifications were attempted by different researchers Kanagaraj et al. (2014b), Jovanovic et al. (2014), Nancharaiah and Mohan (2014) to solve various problems in major areas, such as power energy savings (Piechocki et al. 2014), engineering (Valian and Valian 2014; Esfandiari 2014; Gálvez et al. 2014; Patwardhan et al. 2014; Lim et al. 2014; Fateen and Bonilla-Petriciolet 2014), digital image processing (Bhandari et al. 2014a, b), medical applications (Amsaleka and Latha 2014; Liu and Fu 2014), and clustering and data mining (Cobos et al. 2014; Abbas and Sadeq 2014).

In 2015, CSA underwent remarkable development in different areas in various fields. For instance, in the field of power and energy, Abdelaziz and Ali (2015) used CSA to control the load frequency for a nonlinear interconnected of power system, which achieved better performance in settling times and various indices. The design of an optimum superconducting magnetic energy storage (SMES) system for automatic generation control of a two-area thermal power system using CSA proved the effectiveness of fast-acting energy storage devices, such as SMES (Chaine and Tripathy 2015). Moreover, CSA contributes to energy and cost savings, distributed generation (Kumari and Shukla 2015), plate fin heat exchangers (Wang and Li 2015), short-term hydrothermal scheduling problem (Nguyen et al. 2015, optimal operation of the multi-reservoir system (Ming et al. 2015, and solving economic load dispatch problems (Nguyen and Vo 2015). In the telecommunications and network area, (Ramakrishnan et al. 2015) proposed a design for an adaptive routing protocol using CSA to provide secure and reliable routers between source and destination node, as well as determine an optimal distance and low routing overhead. CSA was used by Nguyen and Truong (2015) to distribute a network reconfiguration for power loss minimization and voltage profile improvement. It was also used by Cao et al. (2015) to discover the transition rules for cellular automata. CSA was also used in the image processing field in Medjahed et al. (2015), Gonzalez et al. (2015), Babu and Sunitha (2015), Biswas et al. (2015), Naik and Panda (2015). Some researchers modified and adapted CSA to enhance performance Li and Yin (2015b), Guerrero et al. (2015), Wang et al. (2015a), increase efficiency Roy et al. (2015),

Wang et al. (2015b), Tran et al. (2015), Li and Yin (2015a), Ding et al. (2015), and reduce time requirement Naik et al. (2015).

In 2016, more than 50 articles used the CSA in various fields. The economic problems of inflation and unfair energy use continue to draw attention toward searching for ways to save energy and money. CSA has a good chance because of its distinctive characteristics; for example, in harnessing wind power through optimal allocation of wind-based distributed generators (Sudabattula and Kowsalya 2016) and reactive power optimization in wind power plants (Pandya et al. 2016), and benefitting from water reservation (Yasar 2016; Nguyen and Vo 2016; Nguyen et al. 2016b) and the optimal design of the power systems (Sanajaoba and Fernandez 2016; Manesh and Ameryan 2016; Elazim and Ali 2016; Devabalaji et al. 2016. Furthermore, CSA was used in power systems to increase the load ability (Abd-Elazim and Ali 2016; Sirjani and Bolan 2016; Nguyen et al. 2016a). Can be noted that the number of publications in engineering field less than number of publications in 2015. CSA has been used in robotic cell scheduling (Majumder and Laha 2016), engineering structure solving (Abdelaziz and Ali 2016; Qu and He 2016; Xu et al. 2016), and electrical engineering (Alssager and Othman 2016). The use of CSA was normal for the rest of the fields, such as economic load dispatch (Tran et al. 2015; Sekhar and Mohanty 2016; Pham et al. 2016), image processing (Zhao et al. 2016; Pare et al. 2016; Raja and Vishnupriya 2016; Suresh and Lal 2016), clustering (Jia et al. 2016) and medical applications (Daniel and Anitha 2016).

Figure 3.3 shows the number of published papers by various researchers from 2009 to 2016 on CSA and its variants based on the different databases where these are

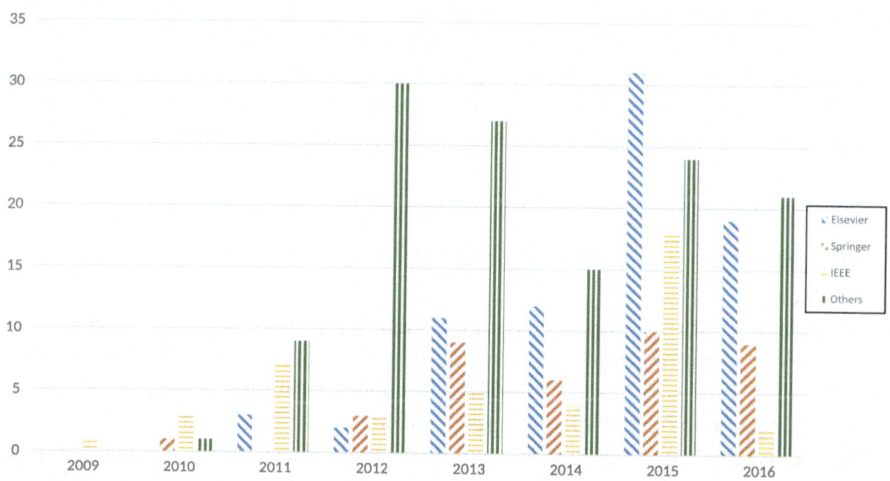

Fig. 3.3 Number of CSA Published Papers By Elsevier, Springer, IEEE and Others, Per Annum (2009 to 2016)

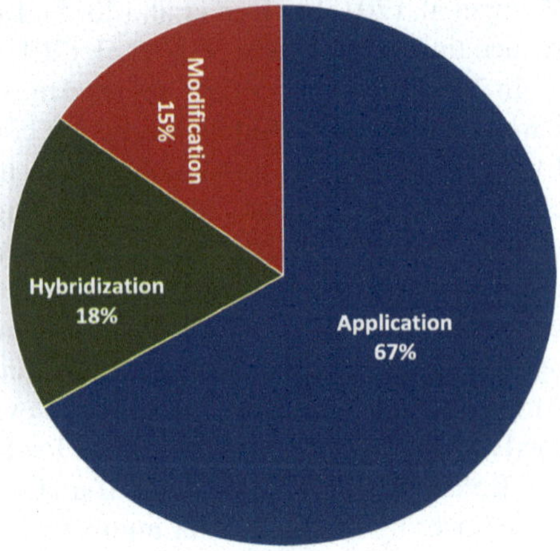

Fig. 3.4 The Distribution of published research articles on CSA

reported; Elsevier, Springer, IEEE, and others.[1] The distribution of published research articles on CSA with respect to applications, hybridizations, and modifications is shown in Fig. 3.4.

3.5 Cuckoo Search Algorithm Variants

The CSA proposed in 2009 is a recent swarm-based algorithm in comparison with the firefly, bee colony, PSO, and ant colony algorithms proposed in 2008, 2005, 1995, and 1992, respectively. However, the CSA has been updated for several variants developed by researchers to cope with the nature of the search space of the optimization problem. Most of these variants will be extensively but not exhaustively described.

3.5.1 Binary Cuckoo Search Algorithm

Gherboudj et al. (2012) proposed a discrete binary cuckoo search (BCS) algorithm for binary optimization problems. The solutions are represented in the optimization problems either by a set of real numbers (called continuous optimization) or by a set of integer numbers (called discrete optimization). The discrete optimization problem class has some subclasses, such as discrete binary optimization problems, and its solutions are represented by a set of bits, including routing (Zhan and Zhang 2009), job shop scheduling (Pongchairerks 2009), and flowshop scheduling problems

[1]This is based on the search conducted by the authors using CSA as the keyword till September, 2016.

(Abualigah et al. 2019). This BCS variation uses a sigmoid function to create a bridge between the discrete/the continuous and the binary values.

As aforementioned, the CSA is based on *Lévy* flights. Therefore, the solutions present as a set of real numbers working in a continuous search space. The solutions must be converted to binary values to extend the CSA to discrete binary areas. The BCS is designed to contain two basic modules:

- Main binary cuckoo dynamics: this module includes two operations:
  - *Lévy* flights: it used to obtain a new cuckoo.
  - Binary solution representation (BSR) to compute the flipping chances for each cuckoo by using the sigmoid function. After that, the flipping chance of each cuckoo to calculate the binary value is used.
- Objective function and the selection operator: the selection operator principles presented here is the same as presented for genetic algorithms

To convert from the continuous area to the binary area, assume x_i is a solution of continuous nature in the interval [0, 1] and $x_i^{'}$ is a BSR, the sigmoid function will convert the values as follows:

$$S(x_i) = \frac{1}{1 + e^{-x_i}} \tag{3.3}$$

where $S(x_i)$ is the flipping chance of bit $x_i^{'}$. To obtain the binary solution $x_i^{'}$, $S(x_i)$ is compared to the result of the generated random number from the [0, 1] interval for each dimension i of solution x as shown in following equation:

$$x_i^{'} = \leftarrow \begin{cases} 1 & \gamma < S(x_i) \\ 0 & otherwise \end{cases}$$

where γ is a random number between [0, 1]. In case the flipping chance of bit $x_i^{'}$ is bigger than the random number then the value is 1, otherwise the value is 0.

3.5.2 Discrete Cuckoo Search Algorithm

The TSP is a classical optimization problem used to evaluate any new development. The TSP principle is that the salesman must visit each city once, starting and finishing from a certain one with a minimum total length of the trip. Ouaarab et al. (2014) introduced a DCSA for the TSP. The author improved and developed the CSA through rebuilding the population and proposing a new category of cuckoos. Thus, DCSA efficiency was increased with less iterations. The DSCA can also solve the continuous and combinatorial problems. It increases the protection of local optima in the case of TSP from stagnation. This supports the DCSA to have more control over the diversification and intensification with less parameters. The experimental analysis of the results showed that the performance of the proposed DCSA algorithm was highly effective compared with genetic simulated annealing ant colony system with particle swarm optimization techniques (GSA-ACS-PSOT) (Chen and Chien 2011) and discrete PSO (Abualigah et al. 2018a).

In another study, DCSA was proposed to solve TSP by Jati et al. (2012). The authors proposed two phases and called them schemes. The first scheme proposed, discrete step size, refers to the distance between the cuckoo and the best cuckoo in its generation. The second scheme is where the cuckoos were updated using a step size α and a random step length drawn from the $L\acute{e}vy$ distribution called $L\acute{e}vy$ flight. The results proved the performance of DCSA with simple TSP. However, it could not achieve the optimal solution for complex TSP. Gherboudj et al. (2012), Shambour (2019) proposed a discrete binary CSA (DBCSA) to solve 0–1 knapsack problems. The authors used a sigmoid function to obtain binary solutions, which are the same as those used in binary PSO. This work has two objectives. The first objective copes with the binary optimization problems, where the basic CSA solution consists of a set of real numbers. On the other hand, the DBCSA solution consists of a set of bits by using a sigmoid function and a probability model to generate binary values. The second objective proves the effectiveness of the basic CSA dealing with binary combinatorial optimization problems. The experimental results on both the multidimensional knapsack problem instances showed the effectiveness of the BDCSA and its ability to obtain good quality solutions compared with the quantum-inspired CSA (QICSA), HS, and binary PSO (Kennedy and Eberhart 1997).

3.5.3 Modified Cuckoo Search Algorithm

Tuba et al. (2011) proposed modified CSA (MCSA) for unconstrained optimization problems. The authors modified the basic CSA by determining the step size from the sorted rather than only the permuted fitness matrix. For example, if the similarity between the cuckoo's egg and the host's eggs was very high, the likelihood of discovery was reduced; therefore, fitness should be linked to the variation in solutions. The authors used a random walk in a biased way with some random step sizes by applying a different function set to calculate this step size. MCSA has been compared with the original CSA by applying a set of standard benchmark functions. The results showed that the performance of MCSA was slightly better in seven of eight benchmark functions.

Abdul Rani et al. (2012) proposed MCSA for side lobe suppression in a symmetric linear antenna array. MCSA was integrated with the roulette wheel and the inertia weight to control the search ability for synthesizing symmetric linear array geometry with the minimum side lobe level. The authors concentrated on the four internal parameters: α, number of population, distribution type, and fraction probability $P\alpha$. In MCSA, when the value of $\alpha = 2.0$, the side lobe level will be lower than the other three counterparts. By contrast, when the value of $\alpha = 1.0$, the side lobe level will approximately have similar performance with the basic CSA of $\alpha = 2.0$. Interpreting these results due to finding the global minimum values require greater probability in the fixed search space. In the distribution type within the direction angle domains of [20°, 75°] and [105°, 160°], MCSA has the best side lobe suppression Abualigah et al. (2018c). The fourth parameter is that the MCSA obtained the best performance with the lowest value of $P\alpha$ ($P\alpha = 0.05$). By contrast, MCSA achieved the worst results when $P\alpha$ had the largest value ($P\alpha = 0.95$). In this work, MCSA was proven to be slightly better than PSO and GA.

The multi-objective analytical aspects of the distribution systems with multiple multi-type compensators solved by MCSA were analyzed (Giridhar et al. 2016). The authors used MCSA to enhance the convergence rate and quality of the solution as well in basic CSA. However, the *Lévy* flight in basic CSA can dominate the exploration of the solutions. MCSA used the crossover operation to try and balance the exploration and exploitation search aspects. This leads to the decrease in diversity, and hence the final best solution with less iterations are achieved. MCSA also generates the initial population with uniform control variables rather than random ones. The effectiveness of MCSA by applying the standard 15-, 33-, and 69-node radial distribution systems seems superior. The obtained results were compared with existing literature methods and achieved more effective performance.

Tawfik et al. (2013) utilized one-rank CSA to optimize algorithmic trading systems. The authors combined the exploration and exploitation phases to generate new solutions. In addition, in basic CSA, exploring new solutions is achieved based on the *Lévy* flights to avoid large moves. Thus, the solutions will skip outside the search space because step size determination depends on the scale of the problem (Shambour et al. 2018). MCSA used the optimal utilization of the *Lévy* flight in optimization and the typical step size $\alpha = 0.01$. MCSA also eliminated the invalid solutions with randomly selected solutions among the current best solutions. MCS was tested on a set of 10 standard benchmark functions and applied to a real-world problem of algorithmic trading systems optimization with clear success. The results showed that the MCSA improved the performance in almost all benchmark functions and other considered problems.

3.5.4 Other Variants of Cuckoo Search Algorithm

The improved CSA (ICSA) was introduced to solve planar graph coloring problem by Zhou et al. (2013). The authors improve the basic CSA by modifying the step length (heavy tailed) through utilizing the walking one strategy, swap and inversion strategy, and greedy transform algorithm. The walking one strategy depends on the node and adjacent nodes of the number of the conflicts. Swapping and inversion strategy finds the maximum conflict node. In this work, the ICSA performance was compared with the basic CSA and the improved POS (IPSO) proposed by Cui et al. (2008). The results showed that ICSA has a fastest convergent velocity and better global search capability in terms of solving the same problem in a more accurate and efficient manner.

Lin et al. (2012) proposed a new CSA variation for the reconstruction of chaotic dynamics, which includes the psychology model effects of emotion and chaotic dynamics. This approach contains a couple of systems; a primary chaotic systems with unknown parameters to generate a time series is used to drive a secondary chaotic systems.

The back propagation neural network algorithm is a widely used technique to train artificial neural networks (Shambour et al. 2019). The Levenberg—Marquardt training algorithm using traditional back propagation optimization has several defects, such as getting stuck in local minima and network stagnation. An improved Levenberg—Marquardt back propagation algorithm integrated and trained with CSA to avoid the local minima problem and achieve fast convergence was proposed by Nawi et al. (2013).

Zhou and Zheng (2013) proposed a new CSA variation based on a complex-valued encoding system. In this algorithm, the term plurality indicates the individual bird's nest. Therefore, a diploid swarm is structured by a sequence plurality, which means that the bird's nest can express the space dimension more than the real-coded one. Thus, the bird's nest contains more information that improves the search capabilities of the pure CSA Abualigah et al. (2018b). Also, the authors used the Gauss distribution with cuckoo search optimization algorithm as a new CSA variation to solve the standard test functions and engineering design optimization problems, called the Gauss cuckoo search (GCS). It achieved better results and a higher convergence rate than CSA alone.

3.6 Previous Works of Hybridization Cuckoo Search Algorithm

A hybrid algorithm is an algorithm that combines two or more other algorithms that solves the different problems. Numerous advantages of hybridization exist, such as enhancing the weaknesses of the algorithm and improving in terms of either computational speed or accuracy. The researchers improved the BCSA performance by integrating the other popular rules and metaheuristic algorithms, such as s selection, mutation, and crossover of GA, and DE. The following subsections present the hybridization of BCSA to other component metaheuristic methods.

3.6.1 *Hybridization of Cuckoo Search Algorithm with Population-Based Algorithms*

A wide range of population-based algorithms, such as DE, PSO, GA, and EA, have been developed to tackle the hard optimization algorithms. In the population-based algorithm behavior, deeply searching in each region of the problem search space is mostly insufficient. Therefore, the search normally does not reach the local optima in that region, especially for the problem of ragged search space. Several suggestions have been proposed to empower the capability of the population based on exploiting the search space region in which it is searching. As commonly known, the operators of the population-based algorithm behaves differently in navigating the search space. Some operators are powerful in exploiting the search space (such as selection in GA), while some others are powerful in exploration (such as random consideration in HS). To this end, hybridizing some powerful operators in a population-based search algorithm may help with striking the right balance between exploration and exploitation through the search.

GA is a search strategy that uses random choice to prove a highly exploitative search by attempting to achieve the balance between exploration and exploitation (Holland 1992). Kanagaraj et al. (2013) proposed a hybridization of BCSA with GA (CSA-GA) to solve reliability and redundancy allocation problems. The solution is achieved through integrating the genetic operators in the basic BCSA. CSA-GA can achieve the balance between exploration and exploitation; therefore, the performance of the algorithm in the search is improved. The crossover preserves the partial identity of the parent cuckoo bird while allowing for the diversity in the solution space. Mutation has been used in the local search of the solution space by making small but logical changes in the genes. There are four classical reliability redundancy allocation problems in the literature that have been applied to confirm the validity of CSA-GA. The results showed that CSA-GA is extremely effective and efficient at locating optimal solutions. Furthermore, it greatly avoids the possibility of trapping in a premature convergence situation.

Kanagaraj et al. (2014a) proved the efficiency of hybrid BCSA and a GA (HCSAGA) for solving engineering design optimization problems. In HCSAGA, it begins with randomly generating the initial solution; thus, it was distributed throughout the solution space. Then, they applied selection, crossover, and mutation, respectively, in each generation. After that, the best solutions were selected by applying some form of elitism. Next, they performed the *Lévy* flight operation for the best solution space. Finally, they replaced the solutions in the current population with better solutions by genetic principles. The previous processes were repeated to guide the search toward the global optimum. To validate the HCSAGA, the authors started with testing 13 standard benchmark functions. Thereafter, they experimented with three well-known design problems reported in the literature. The results proved the performance of HCSAGA compared with the existing approaches for constrained design optimization problems.

The drilling path optimization problem is an example of an NP-hard optimization problem. It becomes an interesting research problem for many researchers because of the complex nature of its search space. The CSGA was proposed to solve the path optimization problem of the drilling process for PCB holes (Kanagaraj et al. 2014b). The authors used genetic principles and replaced it with the BCSA components, such as replacing the crossover with the *Lévy* flights and the mutation with the random walks. CSGA saved the high-quality eggs, while the poor-quality eggs were neglected; this is called the elitism strategy. The experiments showed the efficiency and robustness of the CSGA and its ability to find the optimal path for the PCB holes drilling path optimization problem.

Lim et al. (2016) proposed a hybrid BCSA and a GA for optimizing the hole-making operation (CSAGAHM), knowing that a hole takes time, cost, and effort before obtaining the final size. CSAGAHM minimized the total time of noncutting, tool positioning, and tool switching at the machine. The authors used the same hybridization mechanism used for PCB. CSAGAHM was verified by testing on a set of benchmark problems; the results showed that CSAGAHM is superior in comparison to PSO, immune-based algorithm, ACO, and basic BCSA.

Two hybrid schemes using BCSA and a GA were proposed by Abdel-Baset and Hezam (2016a). There are two phases in both hybrid schemes (i.e., GA-CS and CS-GA). In the first phase, GA or BCSA explores the search space to identify the most promising regions of the search space and find a set of promising solutions. In the second phase, the solution is improved by either GA or BCSA, which did not operate in the first phase. The performances of the GA-CS and CS-GA were tested using 15 benchmark functions carefully selected from the literature. The results illustrated the effectiveness of the proposed algorithms compared with basic BCSA and a GA.

Wang et al. (2012) proposed a hybrid metaheuristic DE and a BCSA (DE/CSA) for the uninhabited combat air vehicle (UCAV) path planning problem and UCAV three-dimension path planning problem. The authors hybridized DE to empower the BCSA capability by embedding the advantages of DE in BCSA. Note that the DE has the skill at exploring the search space and determining the region of global optimal value. Through this DE feature, it can solve the drawback of the basic BCSA, which lies in the speed of exploiting the solution. In addition, DE can add diversity of the population to improve the search efficiency. The performance of the DE/CSA was proved by comparing it with the basic BCSA. The results showed the efficiency and feasibility of DE/CSA to solve both the UCAV path planning problem and UCAV three-dimension path planning problem.

The network operation is decentralized where the nodes implement the network's organization. A mobile ad hoc network (MANET) can connect large networks in a standalone manner. These networks, either connected or standalone, do not need an infrastructure but causes some challenges. The central infrastructure must likewise be ensured by mobile nodes in the new environment. For this purpose, Nancharaiah and Mohan (2014) proposed a hybrid optimization technique, i.e., ACO and a BCSA (ACO-CSA), to optimize MANET routing. ACO-CSA starts to run with ACO to update the pheromone. Then, BCSA is initialized to perform as the local search. The proposed algorithm optimizes the ad hoc on-demand distance vector routing

(AODV) based on delay and cost. The results showed that the performance of the proposed algorithm is better than the basic ACO.

Babukartik and Dhavachelvan (2012) proposed a hybrid BCSA with ACO for the job scheduling problem. In their work, they combined BCSA with some features of ACO. The experimental analysis showed that using the proposed algorithm leads to an increased in the number of tasks. In addition, the time required for task creation and result retrieval is increased.

The optimal design of water distribution systems was solved by using a hybrid optimization method, which combined the BCSA and a HS (CSAHS) (Sheikholeslami et al. 2016). In the CSAHS, the explorative stages of searching and finding the promising regions of the search space were conducted by BCSA. Then, the hybridization between BCSA and HS algorithms was utilized in the exploitation phase. CSAHS was improved through harmony memory (HM), which consists of information extracted online through a search and the addition of the pitch-adjusting operation of HS through the evolution process. Thus, HM vectors become the BCSA population, and an evolving process is performed in the BCSA. In addition, CSAHS was improved by utilizing a self-adaptive strategy for the dynamic adaption of the control parameters using a learning mechanism. The performance of the CSHS was demonstrated by employing four water distribution system design problems with increased scales and complexity. The results illustrated that the CSHS method overtook the basic BCSA, proved its efficiency to obtain optimal solutions, and enhanced the convergence rate.

Wang et al. (2016b) proposed a hybrid algorithm that combined BCSA and a HS (HS/CSA) for continuous optimization problems. In the HS/CSA method, the pitch adjustment of HS was used to update the process of the BCSA, which leads to the increase of population diversity. The improved elitism scheme was used to retain the best individuals in the cuckoo population as well. The performance of HS/CSA was evaluated by means of testing the set of benchmark functions. The obtained results showed that the HS/CSA achieved better outcomes in comparison with ACO, PSO, GA, HS, DE, and basic BCSA.

An early diagnosis of a disease is an important issue in life science and is effective in avoiding the risk to human health. Liu and Fu (2014) proposed a disease diagnosis model based on cuckoo search, PSO, and support vector machine (CSA-POS-SVM). The authors used SVM to solve classification and regression problem on a dataset that is nonlinearly separable and handle data that are not fully linearly separable Reyaz-Ahmed et al. (2009). Additionally, the PSO was used to make use of particles moving in the n-dimensional space to search for solutions to the functions of n-variables. This allows the particles to distribute in the problem space by attracting the particles of the best solutions. Thus, it is initialized with a group of random particles (solutions) and then searches for optima by updating each generation (Ardjani et al. 2010). BCSA discovered that the random-walk style search has better performance for *Lévy* flights rather than the simple random walk. BCSA with its random-walk style is also simple in comparison with other algorithms. Experimental results showed that the CS-PSO-SVM realized better classification accuracy and F-measure than GASVM and PSO-SVM; thus, it is efficient in disease diagnosis.

3.6.2 Hybridization of Cuckoo Search Algorithm with Local Search-Based Algorithm

Population-based search algorithm is normally very powerful in exploring several regions of the problem search space. However, it has difficulty in determining the local optima in each region. By contrast, deep searching of the local search-based algorithm is very efficient in a single search space region but not for several search space regions (Al-Betar et al. 2014). Thus, sometimes, it is very beneficial to hybridize a local and a population search-based method to complement their advantages in a single optimization framework. Based on the above suggestion and through hybridization, the search can strike a balance between the wide range of exploration and nearby exploitation of the problem search space. In this context, BCSA has been hybridized with other population-based algorithms to improve its performance in tackling complex optimization problems. This section summarizes an overview of the attempts made to hybridize BCSA with other population-based algorithms.

Quadratic assignment problems (QAPs) are considered to be NP-hard problems, which cannot be easily solved by exact methods. Therefore, Dejam et al. (2012) proposed a hybrid algorithm combined with the BCSA of TS (i.e., CSA-TS) to solve QAPs. In their research, the QAPs were initially tackled using BCSA. Thereafter, these were combined with TS, which focused on the local search to increase the optimization precision. The experimental results indicated that the proposed algorithm performs better than ABC and GA.

Quantum computing (QC) is a new research field that covers investigations on quantum computers and quantum algorithms (Jaeger 2007). QC can simultaneously deal with huge numbers of quantum states in parallel, enhancing both the efficiency and speed of classical evolutionary algorithms. In this paper, a new inspired algorithm, called the QICSA, for knapsack problem was proposed by Layeb (2011). QICSA was developed to enhance the basic BCSA for successful application on combinatorial optimization problems on some quantum computing principles. In addition, it provides a large exploration of the search space during diversification and intensification stages. The experimental results showed the effectiveness and successful performance of the QICSA on the knapsack problem. Furthermore, it has the ability to achieve good quality solutions. Layeb and Boussalia (2012) used QICSA to solve one-dimensional bin packing problem (1 BPP). In their work, QICSA aimed to determine a suitable quantum representation for the bin packing solutions. A new hybrid quantum measure operation (i.e., using the first-fit heuristic to pack non-filled objects by the standard measure operation) was introduced. The obtained results showed the effectiveness and feasibility of QICSA.

The integration of the Nelder—Mead simplex method with BCSA, named NMA-CSA, was proposed by Jovanovic et al. (2014) to optimize the performance of multi-cell solar systems. In this work, a simplex was used instead of single solutions as nests for the BCSA. In this way, it can replace the Nelder—Mead algorithm instead of the *Lévy*. Thus, NMA-CSA becomes more robust and less sensitive to parameter tuning. However, a difficulty in evaluating the complex function exists.

The performance of NMA-CSA has been applied on both solar cell optimization and standard benchmark functions. The obtained results achieved better solutions with higher speed of convergence compared with other methods presented in the literature. Once again, Ali and Tawhid (2016) proposed a hybrid algorithm, which includes the BCSA and Nelder—Mead method (HCSANM), to solve the integer and minimax optimization problems. HCSANM started with a basic BCSA for a number of iterations, chose the best results achieved, and then passed it to Nelder—Mead, similar to an intensification for increasing the search and eliminating the slow convergence struggle of the basic BCSA. Therefore, HCSANM controls the balance using the basic BCSA for global exploration, and NM method is utilized for deep exploitation. The validation of the performance was determined by applying the HCSANM on 7 integer programming and 10 minimax problems. The results showed the efficiency of HCSANM and its ability to deal with minimax optimization problems in a reasonable time.

The linear least squares problem solved by hybridization algorithm between Newton method (NM) and BCSA is called CSANM (Abdel-Baset and Hezam 2016b). The authors benefited from BCSA for fast convergence and global search as well as from NM for the ability of strong local search. The experimental results showed the convergence efficiency and computational accuracy of the CSANM in comparison with the basic BCSA and HS based on NM (HSNM).

3.6.3 Hybridization of Cuckoo Search Algorithm with Other Techniques

A novel BCSA base on the Gauss distribution (GCSA) was proposed by Zheng and Zhou (2012). In the basic BCSA, although it finds the optimum solution, the search entirely depends on random walks. By contrast, fast convergence and precision cannot be guaranteed. For this purpose, GCSA was introduced to solve the low convergence rate of the basic BCSA. GCSA has been applied to solve the standard test functions and engineering design optimization problems. The obtained results showed that the GCSA proved its efficiency through achieving better solutions compared with basic BCSA.

Chandrasekaran and Simon proposed a hybrid algorithm, which integrated BCSA with the fuzzy system to solve the multi-objective unit commitment problem (Chandrasekaran and Simon 2012). In general, the successful power system needs reliable operation and environment and economic preservation. However, these three points are conflicting because fuel cost is not compatible with the emission and reliability level of the system. The authors used fuzzy set theory to build the fuzzy membership search domain for all possible solutions. They also used BCSA to look for the best solution inside the fuzzy search domain. The proposed method started with the basic BCSA through initializing a fixed number of host nests; each nest includes a cuckoo's egg. The size of the host nest is increased at the end of each generation.

Afterward, the fuzzy fitness starts to pick up the best solution, and then the host bird destroys the alien egg, which is far away from the best solution. Otherwise, the egg grows up alive for the next generation based on the fitness function. The validation of the proposed method is measured using the benchmark test systems for single and multi-objective optimization problems. The results of the proposed method showed its superiority over other methods reported in the literature. It also has the ability to solve non-smooth multi-objective problems in a power system.

Centroidal Voronoi tessellations (CVT) is a Voronoi tessellation of a set where the generators of the Voronoi groups the centers of the same mass where all of them work at the same time (parallel). An integration of BCSA and CVT was proposed to enhance the BCSA (CSA with CVT) (Shatnawi and Nasrudin 2011). The CVT presented the problem space in an equally distributed manner. For this, the authors utilized CVT as the starting points for the nests. Then, the BCSA was used for the generated Voronoi points. The results of the experiment, when compared with the basic BCSA, showed the effectiveness, robustness, and feasibility of the BCSA with CVT.

3.7 Conclusion

In this chapter, over 150 research articles were studied and analyzed in an attempt to make a robust conclusion for the researchers interested in working on CSA. The review exhaustively and comprehensively summarized the references published from 2009 until the second half of 2016. The papers, which were collected in this work, observed that the largest share went to CSA applications in various fields, such as medical, image processing, economic load dispatch, engineering design, and data clustering. This result shows the high popularity of CSA because of its successful characteristics and features. The modification and three categories of hybridizations were introduced in this chapter to enhance the performance of CSA and maintain its applicability in the problems or a rugged search space.

CSA remains a promising and interesting algorithm and will continue to be widely used by researchers across different fields. Its advantages over other optimization algorithms include its simplicity, fewer parameters compared with other algorithms, and ease of hybridization with other optimization algorithms. However, CSA lacks the mathematical analysis. It does not have a theoretical analysis similar to other algorithms, such as PSO (Schmitt 2015) and GA (Wróblewski 1996). This difference can be observed with the difficulty of understanding when and why the algorithm works. Additionally, how does the performance of the algorithm improve compared to other search techniques parameter tuning is also considered an important part of research (Eiben and Smit 2011), where the values and settings of the parameters govern the CSA performance.

All these reasons make CSA robustly viable for continued utilization to the community. This chapter could also guide the researchers who are currently working or will work in this area by leading them toward how the CSA algorithm can be employed to solve the problems, point out its weaknesses and strengths, and prove its effectiveness. Thus, important research can still be solved through the utilization of CSA.

References

Abbas, A. K., & Sadeq, A. T. (2014). Database clustering using intelligent techniques.

Abd Elazim, S. M., & Ali, E. S. (2016). Optimal power system stabilizers design via cuckoo search algorithm. *International Journal of Electrical Power & Energy Systems, 75*, 99–107.

Abd-Elazim, S. M., & Ali, E. S. (2016). Optimal location of statcom in multimachine power system for increasing loadability by cuckoo search algorithm. *International Journal of Electrical Power & Energy Systems, 80*, 240–251.

Abdelaziz, A. Y., & Ali, E. S. (2016). Load frequency controller design via artificial cuckoo search algorithm. *Electric Power Components and Systems, 44*(1), 90–98.

Abdelaziz, A. Y., & Ali, E. S. (2015). Cuckoo search algorithm based load frequency controller design for nonlinear interconnected power system. *International Journal of Electrical Power & Energy Systems, 73*, 632–643.

Abdel-Baset, M., & Hezam, I. M. (2016b). Solving linear least squares problems based on improved cuckoo search algorithm. *Mathematical sciences, 5*(2), 199–202.

Abdel-Baset, M., & Hezam, I. (2016a). Cuckoo search and genetic algorithm hybrid schemes for optimization problems. *Applied Mathematics, 10*(3), 1185–1192.

Abdul Rani, K. N., Abdul Malek, M. F., & Siew-Chin, N. (2012). Nature-inspired cuckoo search algorithm for side lobe suppression in a symmetric linear antenna array. *Radioengineering*.

Abualigah, L. M., Khader, A. T., & Hanandeh, E. S. (2018a). A combination of objective functions and hybrid krill herd algorithm for text document clustering analysis. *Engineering Applications of Artificial Intelligence, 73*, 111–125.

Abualigah, L. M., Khader, A. T., & Hanandeh, E. S. (2018b). Hybrid clustering analysis using improved krill herd algorithm. *Applied Intelligence*, pp. 1–25.

Abualigah, L. M., Khader, A. T., & Hanandeh, E. S. (2018c). A new feature selection method to improve the document clustering using particle swarm optimization algorithm. *Journal of Computational Science, 25*, 456–466.

Abualigah, L. M., Khader, A. T., & Hanandeh, E. S. (2019). Modified krill herd algorithm for global numerical optimization problems. In *Advances in Nature-Inspired Computing and Applications*, pp. 205–221. Springer.

Abualigah, L. M., Sawaie, A. M., Khader, A. T., Rashaideh, H., Al-Betar, M. A., & Shehab, M. (2017b). β-hill climbing technique for the text document clustering. *New Trends in Information Technology, 60*.

Agrawal, S., Panda, R., Bhuyan, S., & Panigrahi, B. K. (2013). Tsallis entropy based optimal multilevel thresholding using cuckoo search algorithm. *Swarm and Evolutionary Computation, 11*, 16–30.

Ahmed, J., & Salam, Z. (2013). A soft computing mppt for pv system based on cuckoo search algorithm. In *Power Engineering, Energy and Electrical Drives (POWERENG), 2013 Fourth International Conference on*, pp. 558–562. IEEE.

Ahmed, J., & Salam, Z. (2014). A maximum power point tracking (mppt) for pv system using cuckoo search with partial shading capability. *Applied Energy, 119*, 118–130.

Al-Betar, M. A., Khader, A. T., & Doush, I. A. (2014). Memetic techniques for examination timetabling. *Annals of Operations Research, 218*(1), 23–50.

Ali, A. F., & Tawhid, M. A. (2016). A hybrid cuckoo search algorithm with nelder mead method for solving global optimization problems. *SpringerPlus, 5*(1): 1.

Alssager, M., & Othman, Z. A. (2016). Taguchi-based parameter setting of cuckoo search algorithm for capacitated vehicle routing problem. In *Advances in Machine Learning and Signal Processing*, pp. 71–79. Springer.

Amsaleka, R., & Latha, M. (2014). A optimally enhanced fuzzy kc means (oefkcm) for clustering algorithm medical image segmentation. *Work*, *3*(3).

Ardjani, F., Sadouni, K., & Benyettou, M. (2010). Optimization of svm multiclass by particle swarm (pso-svm). In *2010 2nd International Workshop on Database Technology and Applications*, pp. 1–4. IEEE.

Babu, R. K., & Sunitha, K. V. N. (2015). Enhancing digital images through cuckoo search algorithm in combination with morphological operation. *Journal of Computer Science*, *11*(1), 7.

Babukartik, R. G., & Dhavachelvan, P. (2012). Hybrid algorithm using the advantage of aco and cuckoo search for job scheduling. *International Journal of Information Technology Convergence and Services*, *2*(4), 25.

Basu, M., & Chowdhury, A. (2013). Cuckoo search algorithm for economic dispatch. *Energy*, *60*, 99–108.

Bhandari, A. K., Singh, V. K., Kumar, A., & Singh, G. K. (2014b). Cuckoo search algorithm and wind driven optimization based study of satellite image segmentation for multilevel thresholding using kapur x92s entropy. *Expert Systems with Applications*, *41*(7), 3538–3560.

Bhandari, A. K., Soni, V., Kumar, A., & Singh, G. K. (2014a). Cuckoo search algorithm based satellite image contrast and brightness enhancement using dwt-svd. *ISA Transactions*, *53*(4), 1286–1296.

Bhargava, V., Fateen, S.-E. K., & Bonilla-Petriciolet, A. (2013). Cuckoo search: a new nature-inspired optimization method for phase equilibrium calculations. *Fluid Phase Equilibria*, *337*, 191–200.

Biswas, B., Roy, P., Choudhuri, R., & Sen, B. K. (2015). Microscopic image contrast and brightness enhancement using multi-scale retinex and cuckoo search algorithm. *Procedia Computer Science*, *70*, 348–354.

Bolaji, A.L., Al-Betar, M. A., Awadallah, M. A., Khader, A. T., Abualigah, L. M. (2016). A comprehensive review: Krill herd algorithm (kh) and its applications. *Applied Soft Computing*, *49*, 437–446.

Brown, C. T., Liebovitch, L. S., & Glendon, R. (2007). Lévy flights in dobe ju/x92hoansi foraging patterns. *Human Ecology*, *35*(1), 129–138.

Buaklee, W., & Hongesombut, K. (2013). Optimal dg allocation in a smart distribution grid using cuckoo search algorithm. In *Electrical Engineering/Electronics, Computer, Telecommunications and Information Technology (ECTI-CON), 2013 10th International Conference on*, pp. 1–6. IEEE.

Bulatović, R. R., DJordjević, S. R., & DJordjević, V. S. (2013). Cuckoo search algorithm: a metaheuristic approach to solving the problem of optimum synthesis of a six-bar double dwell linkage. *Mechanism and Machine Theory*, *61*, 1–13.

Cao, M., Tang, G., Shen, Q., & Wang, Y. (2015). A new discovery of transition rules for cellular automata by using cuckoo search algorithm. *International Journal of Geographical Information Science*, (ahead-of-print), 1–19.

Chaine, S., & Tripathy, M. Design of an optimal smes for automatic generation control of two-area thermal power system using cuckoo search algorithm. *Journal of Electrical Systems and Information Technology*.

Chen, S.-M., & Chien, C.-Y. (2011). Solving the traveling salesman problem based on the genetic simulated annealing ant colony system with particle swarm optimization techniques. *Expert Systems with Applications*, *38*(12), 14439–14450.

Cobos, C., Muñoz-Collazos, H., Urbano-Muñoz, R., Mendoza, M., León, E., & Herrera-Viedma, E. (2014). Clustering of web search results based on the cuckoo search algorithm and balanced bayesian information criterion. *Information Sciences*, *281*, 248–264.

Cui, G., Qin, L., Liu, S., Wang, Y., Zhang, X., & Cao, X. (2008). Modified pso algorithm for solving planar graph coloring problem. *Progress in Natural Science*, *18*(3), 353–357.

Daniel, E., & Anitha, J. (2016). Optimum wavelet based masking for the contrast enhancement of medical images using enhanced cuckoo search algorithm. *Computers in Biology and Medicine*, *71*, 149–155.

Davies, G. H. (1970). The life of birds, parenthood. http://www.pbs.org/lifeofbirds/home/index.html.

Dejam, S., Sadeghzadeh, M., & Mirabedini, S. J. (2012). Combining cuckoo and tabu algorithms for solving quadratic assignment problems. *Journal of Academic and Applied Studies*, *2*(12), 1–8.

Devabalaji, K. R., Yuvaraj, T., & Ravi, K. (2016). An efficient method for solving the optimal sitting and sizing problem of capacitor banks based on cuckoo search algorithm. *Ain Shams Engineering Journal*.

Ding, X., Xu, Z., Cheung, N. J., & Liu, X. (2015). Parameter estimation of takagi–sugeno fuzzy system using heterogeneous cuckoo search algorithm. *Neurocomputing*, 151, 1332–1342.

Durgun, İ., & Yildiz, A. R. (2012). Structural design optimization of vehicle components using cuckoo search algorithm. *Materials Testing*, *54*(3), 185–188.

Eiben, A. E., & Smit, S. K. (2011). Parameter tuning for configuring and analyzing evolutionary algorithms. *Swarm and Evolutionary Computation*, *1*(1), 19–31.

Esfandiari, A. (2014). Cuckoo optimization algorithm in cutting conditions during machining. *Journal of Advances in Computer Research*, *5*(2), 45–57.

Fateen, S.-E. K., & Bonilla-Petriciolet, A. (2014). Unconstrained gibbs free energy minimization for phase equilibrium calculations in nonreactive systems, using an improved cuckoo search algorithm. *Industrial & Engineering Chemistry Research*, *53*(26), 10826–10834.

Femia, N., Petrone, G., Spagnuolo, G., & Vitelli, M. (2005). Optimization of perturb and observe maximum power point tracking method. *IEEE Transactions on Power Electronics*, *20*(4), 963–973.

Fister Jr, I., Yang, X.-S., Fister, D., & Fister, I. (2014). Cuckoo search: a brief literature review. In *Cuckoo Search and Firefly Algorithm*, pp. 49–62. Springer.

Gálvez, A., Iglesias, A., & Cabellos, L. (2014). Cuckoo search with lévy flights for weighted bayesian energy functional optimization in global-support curve data fitting. *The Scientific World Journal*.

Gandomi, A. H., Talatahari, S., Yang, X.-S., & Deb, S. (2013a). Design optimization of truss structures using cuckoo search algorithm. *The Structural Design of Tall and Special Buildings*, *22*(17), 1330–1349.

Gandomi, A. H., Yang, X.-S., & Alavi, A. H. (2013b). Cuckoo search algorithm: a metaheuristic approach to solve structural optimization problems. *Engineering with Computers*, *29*(1), 17–35.

Geem, Z. W., Kim, J. H., & Loganathan, G. V. (2001). A new heuristic optimization algorithm: Harmony search. *Simulation*, *76*(2), 60–68.

Gherboudj, A., Layeb, A., & Chikhi, S. (2012). Solving 0–1 knapsack problems by a discrete binary version of cuckoo search algorithm. *International Journal of Bio-Inspired Computation*, *4*(4), 229–236.

Giridhar, M. S., Sivanagaraju, S., Suresh, C. V., & Umapathi Reddy, P. (2016). Analyzing the multi objective analytical aspects of distribution systems with multiple multi-type compensators using modified cuckoo search algorithm. *International Journal of Parallel, Emergent and Distributed Systems*, pp. 1–23.

Giveki, D., Salimi, H., Bahmanyar, G., & Khademian, Y. (2012). Automatic detection of diabetes diagnosis using feature weighted support vector machines based on mutual information and modified cuckoo search. arXiv preprint arXiv:1201.2173.

Glover, F. (1977). Heuristics for integer programming using surrogate constraints. *Decision Sciences*, *8*(1), 156–166.

Goel, S., Sharma, A., & Bedi, P. (2011). Cuckoo search clustering algorithm: A novel strategy of biomimicry. In *Information and Communication Technologies (WICT), 2011 World Congress on*, pp. 916–921. IEEE.

Gonzalez, C. I., Castro, J. R., Melin, P., & Castillo, O. (2015). Cuckoo search algorithm for the optimization of type-2 fuzzy image edge detection systems. In *Evolutionary Computation (CEC), 2015 IEEE Congress on*, pp. 449–455. IEEE.

Guerrero, M., Castillo, O., & Garcia, M. (2015). Fuzzy dynamic parameters adaptation in the cuckoo search algorithm using fuzzy logic. In *Evolutionary Computation (CEC), 2015 IEEE Congress on*, pp. 441–448. IEEE.

Holland, J. H. (1975). *Adaptation in natural and artificial systems: An introductory analysis with applications to biology, control, and artificial intelligence*. U Michigan Press.

Holland, J. H. (1992). *Adaptation in natural and artificial systems: An introductory analysis with applications to biology, control, and artificial intelligence*. MIT Press.

Hybrid approach using fuzzy assisted cuckoo search algorithm. (2012). K Chandrasekaran and Sishaj P Simon. Multi-objective scheduling problem. *Swarm and Evolutionary Computation, 5*, 1–16.

Jaeger, G. (2007). *Quantum information*. Springer.

James, K., & Russell, E. (1995). Particle swarm optimization. In *Proceedings of 1995 IEEE International Conference on Neural Networks*, pp. 1942–1948.

Jati, G. K., Manurung, H. M., et al. (2012). Discrete cuckoo search for traveling salesman problem. In *Computing and Convergence Technology (ICCCT), 2012 7th International Conference on*, pp. 993–997. IEEE.

Jia, B., Biting, Y., Qi, W., Yang, X., Wei, C., Law, R., et al. (2016). Hybrid local diffusion maps and improved cuckoo search algorithm for multiclass dataset analysis. *Neurocomputing, 189*, 106–116.

Jovanovic, R., Kais, S., & Alharbi, F. H. (2014). Cuckoo search inspired hybridization of the nelder-mead simplex algorithm applied to optimization of photovoltaic cells. arXiv preprint arXiv:1411.0217.

Kanagaraj, G., Ponnambalam, S. G., & Jawahar, N. (2013). A hybrid cuckoo search and genetic algorithm for reliability–redundancy allocation problems. *Computers & Industrial Engineering, 66*(4), 1115–1124.

Kanagaraj, G., Ponnambalam, S. G., & Lim, W. C. E. (2014b). Application of a hybridized cuckoo search-genetic algorithm to path optimization for pcb holes drilling process. In *Automation Science and Engineering (CASE), 2014 IEEE International Conference on*, pp. 373–378. IEEE.

Kanagaraj, G., Ponnambalam, S. G., Jawahar, N., & Nilakantan, J. M. (2014a). An effective hybrid cuckoo search and genetic algorithm for constrained engineering design optimization. *Engineering Optimization, 46*(10), 1331–1351.

Karaboga, D. (2005). An idea based on honey bee swarm for numerical optimization. Technical report, Technical report-tr06, Erciyes university, engineering faculty, computer engineering department.

Kaveh, A., & Bakhshpoori, T. (2016). An efficient multi-objective cuckoo search algorithm for design optimization. *Advances in Computational Design, 1*(1), 87–103.

Kennedy, J., & Eberhart, R. C. (1997). A discrete binary version of the particle swarm algorithm. In *Systems, Man, and Cybernetics, 1997. Computational Cybernetics and Simulation, 1997 IEEE International Conference on*, volume 5, pp. 4104–4108. IEEE.

Khan, K., & Sahai, A. (2013). Neural-based cuckoo search of employee health and safety (hs). *International Journal of Intelligent Systems and Applications, 5*(2), 76.

Kirkpatrick, S., Gelatt, C. D., Vecchi, M. P., et al. (1983). Optimization by simmulated annealing. *Science, 220*(4598), 671–680.

Koza, J. R. (1994). Genetic programming ii: Automatic discovery of reusable subprograms. Cambridge, MA, USA.

Koziel, S. & Yang, X.-S. (2011). *Computational optimization, methods and algorithms* (Vol. 356). Springer.

Kumari, A., & Shukla, S. (2015). Distributed generation allocation and voltage improvement in distribution system using cuckoo search algorithm. *International Journal of Engineering Science and Technology, 7*(9), 298.

Layeb, A.., & Boussalia, S. R. (2012). A novel quantum inspired cuckoo search algorithm for bin packing problem. *International Journal of Information Technology and Computer Science (IJITCS)*, *4*(5), 58.

Layeb, A. (2011). A novel quantum inspired cuckoo search for knapsack problems. *International Journal of bio-inspired Computation*, *3*(5), 297–305.

Li, X., & Yin, M. (2015b). A particle swarm inspired cuckoo search algorithm for real parameter optimization. *Soft Computing*, pp. 1–25.

Li, X., & Yin, M. (2015a). Modified cuckoo search algorithm with self adaptive parameter method. *Information Sciences*, *298*, 80–97.

Li, X., & Yin, M. (2016). A particle swarm inspired cuckoo search algorithm for real parameter optimization. *Soft Computing*, *20*(4), 1389–1413.

Lim, W. C. E., Kanagaraj, G., & Ponnambalam, S. G. (2014). Pcb drill path optimization by combinatorial cuckoo search algorithm. *The Scientific World Journal*.

Lim, W. C. E., Kanagaraj, G., & Ponnambalam, S. G. (2016). A hybrid cuckoo search-genetic algorithm for hole-making sequence optimization. *Journal of Intelligent Manufacturing*, *27*(2), 417–429.

Lin, J.-H., Lee, I.-H., et al. (2012). Emotional chaotic cuckoo search for the reconstruction of chaotic dynamics. In *source: 11th WSEAS Int. Conf. on Computational Intelligence, Man-Machine Systems and Cybernetics (CIMMACS'12)*, pp. 123–128.

Liu, X. & Hui, F. (2014). Pso-based support vector machine with cuckoo search technique for clinical disease diagnoses. *The Scientific World Journal*.

Ma, J., Ting, T. O., Man, K. L., Zhang, N., Guan, S.-U., & Wong, P. W. H. (2013). Parameter estimation of photovoltaic models via cuckoo search. *Journal of Applied Mathematics*, *2013*.

Machowski, J., Bialek, J., & Bumby, J. (2011). *Power system dynamics: Stability and control*. Wiley.

Majumder, A., & Laha, D. (2016). A new cuckoo search algorithm for 2-machine robotic cell scheduling problem with sequence-dependent setup times. *Swarm and Evolutionary Computation*, *28*, 131–143.

Manesh, M. H. K., & Ameryan, M. (2016). Optimal design of a solar-hybrid cogeneration cycle using cuckoo search algorithm. *Applied Thermal Engineering*, *102*, 1300–1313.

Medjahed, S. A., Saadi, T. A., Benyettou, A., & Ouali, M. (2015). Binary cuckoo search algorithm for band selection in hyperspectral image classification. *IAENG International Journal of Computer Science*, *42*(3).

Ming, B., Chang, J., Huang, Q., Wang, Y., & Huang, S. (2015). Optimal operation of multi-reservoir system based-on cuckoo search algorithm. *Water Resources Management*, *29*(15), 5671–5687.

Naik, M. K., & Panda, R. (2015). A novel adaptive cuckoo search algorithm for intrinsic discriminant analysis based face recognition. *Applied Soft Computing*.

Naik, M., Nath, M. R., Wunnava, A., Sahany, S., & Panda, R. (2015). A new adaptive cuckoo search algorithm. In *Recent Trends in Information Systems (ReTIS), 2015 IEEE 2nd International Conference on*, pp. 1–5. IEEE.

Nancharaiah, B., & Mohan, B. C. (2014). Hybrid optimization using ant colony optimization and cuckoo search in manet routing. In *Communications and Signal Processing (ICCSP), 2014 International Conference on*, pp. 1729–1734. IEEE.

Nawi, N. M., Khan, A., & Rehman, M. Z. (2013). A new cuckoo search based levenberg-marquardt (cslm) algorithm. In *Computational Science and Its Applications—ICCSA 2013*, pp. 438–451. Springer.

Nguyen, T. T., & Truong, A. V. (2015). Distribution network reconfiguration for power loss minimization and voltage profile improvement using cuckoo search algorithm. *International Journal of Electrical Power & Energy Systems*, *68*, 233–242.

Nguyen, T. T., & Vo, D. N. (2015). The application of one rank cuckoo search algorithm for solving economic load dispatch problems. *Applied Soft Computing*, *37*, 763–773.

Nguyen, T. T., & Vo, D. N. (2016). Solving short-term cascaded hydrothermal scheduling problem using modified cuckoo search algorithm. *International Journal of Grid and Distributed Computing*, *9*(1), 67–78.

Nguyen, K. P., Fujita, G., & Dieu, V. N. (2016a). Cuckoo search algorithm for optimal placement and sizing of static var compensator in large-scale power systems. *Journal of Artificial Intelligence and Soft Computing Research*, *6*(2), 59–68.

Nguyen, T. T., Vo, D. N., & Dinh, B. H. (2016b). Cuckoo search algorithm using different distributions for short-term hydrothermal scheduling with reservoir volume constraint. *International Journal on Electrical Engineering and Informatics*, *8*(1), 76.

Nguyen, T. T., Vo, D. N., & Ongsakul, W. (2015). One rank cuckoo search algorithm for short-term hydrothermal scheduling with reservoir constraint. In *PowerTech, 2015 IEEE Eindhoven*, pp. 1–6. IEEE.

Noghrehabadi, A., Ghalambaz, M., Ghalambaz, M., & Vosough, A. (2011). A hybrid power series x97cuckoo search optimization algorithm to electrostatic deflection of micro fixed-fixed actuators. *International Journal of Multidisciplinary Sciences and Engineering*, *2*(4), 22–26.

Ong, P., & Kohshelan, S. (2016). Performances of adaptive cuckoo search algorithm in engineering optimization.

Ouaarab, A., Ahiod, B., & Yang, X.-S. (2014). Discrete cuckoo search algorithm for the travelling salesman problem. *Neural Computing and Applications*, *24*(7–8), 1659–1669.

Pandya, K. S., Pandya, J. K., Joshi, S. K., & Mewada, H. K. (2016). Reactive power optimization in wind power plants using cuckoo search algorithm. In *Metaheuristics and Optimization in Civil Engineering*, pp. 181–197. Springer.

Pare, S., Kumar, A., Bajaj, V., & Singh, G. K. (2016). A multilevel color image segmentation technique based on cuckoo search algorithm and energy curve. *Applied Soft Computing*, *47*, 76–102.

Patwardhan, A. P., Patidar, R., & George, N. V. (2014). On a cuckoo search optimization approach towards feedback system identification. *Digital Signal Processing*, *32*, 156–163.

Pavlyukevich, I. (2007). Lévy flights, non-local search and simulated annealing. *Journal of Computational Physics*, *226*(2), 1830–1844.

Pham, L. H., Nguyen, T. T., Vo, D. N., & Tran, C. D. (2016). Adaptive cuckoo search algorithm based method for economic load dispatch with multiple fuel options and valve point effect. *Fuel*, *9*(1).

Piechocki, J., Ambroziak, D., Palkowski, A., & Redlarski, G. (2014). Use of modified cuckoo search algorithm in the design process of integrated power systems for modern and energy self-sufficient farms. *Applied Energy*, *114*, 901–908.

Pongchairerks, P. (2009). Particle swarm optimization algorithm applied to scheduling problems. *ScienceAsia*, *35*(1), 89–94.

Qu, C., & He, W. (2016). A cuckoo search algorithm with complex local search method for solving engineering structural optimization problem. In *MATEC Web of Conferences*, vol. 40. EDP Sciences.

Rajabioun, R. (2011). Cuckoo optimization algorithm. *Applied Soft Computing*, *11*(8), 5508–5518.

Ramakrishnan, B., Sreedivya, S. R., & Selvi, M. (2015). Adaptive routing protocol based on cuckoo search algorithm (arp-cs) for secured vehicular ad hoc network (vanet). *International Journal of Computer Networks and Applications (IJCNA)*, *2*(4), 173–178.

Reyaz-Ahmed, A., Zhang, Y.-Q., & Harrison, R. W. (2009). Granular decision tree and evolutionary neural svm for protein secondary structure prediction. *International Journal of Computational Intelligence Systems*, *2*(4), 343–352.

Roy, S., Mallick, A., Chowdhury, S. S., & Roy, S. (2015). A novel approach on cuckoo search algorithm using gamma distribution. In *Electronics and Communication Systems (ICECS), 2015 2nd International Conference on*, pp. 466–468. IEEE.

Roy, S., & Chaudhuri, S. (2013). Cuckoo search algorithm using lévy flight: A review. *International Journal of Modern Education and Computer Science (IJMECS)*, *5*(12), 10.

Sanajaoba, S., & Fernandez, E. (2016). Maiden application of cuckoo search algorithm for optimal sizing of a remote hybrid renewable energy system. *Renewable Energy*, *96*, 1–10.

Schmitt, B. I. (2015). *Convergence analysis for particle swarm optimization*. FAU University Press.

Sekhar, P., & Mohanty, S. (2016). An enhanced cuckoo search algorithm based contingency constrained economic load dispatch for security enhancement. *International Journal of Electrical Power & Energy Systems, 75*, 303–310.

Shambour, M. K., & Yousef. (2019). Adaptive multi-crossover evolutionary algorithm for real-world optimisation problems. *International Journal of Reasoning-based Intelligent Systems, 11*(1), 1–10.

Shambour, M. K. Y., Abusnaina, A. A., & Alsalibi, A. I. (2019). Modified global flower pollination algorithm and its application for optimization problems. *Interdisciplinary sciences, computational life sciences, 11*(3), 496–507.

Shambour, Y., et al. (2018). Vibrant search mechanism for numerical optimization functions. *Journal of Information & Communication Technology, 17*(4).

Shatnawi, M., & Nasrudin, M. F. (2011). Starting configuration of cuckoo search algorithm using centroidal voronoi tessellations. In *Hybrid Intelligent Systems (HIS), 2011 11th International Conference on*, pages 40–45. IEEE.

Shehab, M., & Khader, A. T. (2018). *Modified cuckoo search algorithm using a new selection scheme for unconstrained optimization problems. 14*, 1.

Shehab, M., Daoud, M. Sh., AlMimi, H. M., Abualigah, L. M., & Khader, A. T. (2019a). Hybridizing cuckoo search algorithm for extracting the odf maxima in spherical harmonic representation. *International Journal of Bio-Inspired Computation*, (in press).

Shehab, M., Khader, A. T., & Al-Betar, M. A. (2017a). A survey on applications and variants of the cuckoo search algorithm. *Applied Soft Computing*.

Shehab, M., Khader, A. T., & Al-Betar, M. A., New selection schemes for particle swarm optimization. *IEEJ Transactions on Electronics, Information and Systems, 136*(12), 1706–1711. https://doi.org/10.1541/ieejeiss.136.1706.

Shehab, M., Khader, A. T., & Alia, M. A. (2019b). Enhancing cuckoo search algorithm by using reinforcement learning for constrained engineering optimization problems. In *2019 IEEE Jordan International Joint Conference on Electrical Engineering and Information Technology (JEEIT)*, pp. 812–816. IEEE.

Shehab, M., Khader, A. T., & Laouchedi, M. (2017c). Modified cuckoo search algorithm for solving global optimization problems. In *International Conference of Reliable Information and Communication Technology*, pp. 561–570. Springer.

Shehab, M., Khader, A. T., & Laouchedi, M. (2018a). A hybrid method based on cuckoo search algorithm for global optimization problems. *Journal of ICT, 17*(3), 469–491.

Shehab, M., Khader, A. T., Al-Betar, M. A., & Abualigah, L. M. (2017b). Hybridizing cuckoo search algorithm with hill climbing for numerical optimization problems. In *Information Technology (ICIT), 2017 8th International Conference on*, pp. 36–43. IEEE.

Shehab, M., Khader, A. T., Laouchedi, M., & Alomari, O. A. (2018b). Hybridizing cuckoo search algorithm with bat algorithm for global numerical optimization. *The Journal of Supercomputing*, 1–28.

Sheikholeslami, R., Zecchin, A. C., Zheng, F., Talatahari, S. (2016). A hybrid cuckoo–harmony search algorithm for optimal design of water distribution systems. *Journal of Hydroinformatics, 18*(3): 544–563.

Sirjani, R., & Bolan, N. T. (2016). An improved cuckoo search algorithm for voltage stability enhancement in power transmission networks. *World Academy of Science, Engineering and Technology, International Journal of Electrical, Computer, Energetic, Electronic and Communication Engineering, 10*(5), 513–517.

Sri Madhava Raja, N., & Vishnupriya, R. (2016). Kapurx92s entropy and cuckoo search algorithm assisted segmentation and analysis of rgb images. *Indian Journal of Science and Technology, 9*(17).

Stewart, B., Wild, C. P. et al. (2016). World cancer report 2014. *World*.

Storn, R., & Price, K. V. (1996). Minimizing the real functions of the icec'96 contest by differential evolution. In *International Conference on Evolutionary Computation*, pp. 842–844.

Sudabattula, S., & Kowsalya, M. (2016). Optimal allocation of wind based distributed generators in distribution system using cuckoo search algorithm. *Procedia Computer Science*, *92*, 298–304.

Suresh, S., & Lal, S. (2016). An efficient cuckoo search algorithm based multilevel thresholding for segmentation of satellite images using different objective functions. *Expert Systems with Applications*, *58*, 184–209.

Talatahari, S., Rahbari, N. M., & Kaveh, A. (2013). A new hybrid optimization algorithm for recognition of hysteretic non-linear systems. *KSCE Journal of Civil Engineering*, *17*(5), 1099–1108.

Tawfik, A. S., Badr, A. A., & Abdel-Rahman, I. F. (2013). One rank cuckoo search algorithm with application to algorithmic trading systems optimization. *International Journal of Computer Applications*, *64*(6).

Tiwari, V. (2012). Face recognition based on cuckoo search algorithm. *Image*, *7*(8), 9.

Tran, C. D., Dao, T. T., Vo, V. S., & Nguyen, T. T. (2015). Economic load dispatch with multiple fuel options and valve point effect using cuckoo search algorithm with different distributions. *International Journal of Hybrid Information Technology*, *8*(1), 305–316.

Tuba, M., Subotic, M., & Stanarevic, N. (2011). Modified cuckoo search algorithm for unconstrained optimization problems. In *Proceedings of the 5th European Conference on European Computing Conference*, pp. 263–268. World Scientific and Engineering Academy and Society (WSEAS).

Valian, E., & Valian, E. (2014). A cuckoo search algorithm by lévy flights for solving reliability redundancy allocation problems. *Engineering Optimization*, *45*(11), 1273–1286.

Valian, E., Mohanna, S., & Tavakoli, S. (2011). Improved cuckoo search algorithm for feedforward neural network training. *International Journal of Artificial Intelligence & Applications*, *2*(3), 36–43.

Viswanathan, G. M., Bartumeus, F., Buldyrev, S. V., Catalan, J., Fulco, U. L., Havlin, S., Da Luz, M. G. E. , Lyra, M. L., Raposo, E. P., & Stanley, H. E. (2002). Lévy flight random searches in biological phenomena. *Physica A: Statistical Mechanics and its Applications*, *314*(1), 208–213.

Viswanathan, G. M., Buldyrev, S. V., Havlin, S., Da Luz, M. G. E., Raposo, E. P., & Stanley, H. E. (1999). Optimizing the success of random searches. *Nature*, *401*(6756), 911–914.

Vo, D. N., Schegner, P., & Ongsakul, W. (2013). Cuckoo search algorithm for non-convex economic dispatch. *IET Generation, Transmission & Distribution*. *7*(6), 645–654.

Walton, S., Hassan, O., Morgan, K., & Brown, M. R. (2011b). Modified cuckoo search: a new gradient free optimisation algorithm. *Chaos, Solitons & Fractals*, *44*(9), 710–718.

Wang, G.-G., Gandomi, A. H., Zhao, X., & Chu, H. C. E. (2016b). Hybridizing harmony search algorithm with cuckoo search for global numerical optimization. *Soft Computing*, *20*(1), 273–285.

Wang, J.-S., Li, S.-X., & Song, J.-D. (2015b). Cuckoo search algorithm based on repeat-cycle asymptotic self-learning and self-evolving disturbance for function optimization. *Computational Intelligence and Neuroscience*.

Wang, H., Wang, W., Sun, H., Li, C., Rahnamayan, S., & Liu ,Y. (2015a). A modified cuckoo search algorithm for flow shop scheduling problem with blocking. In *Evolutionary Computation (CEC), 2015 IEEE Congress on*, pp. 456–463. IEEE.

Wang, Z., & Li, Y. (2015). Irreversibility analysis for optimization design of plate fin heat exchangers using a multi-objective cuckoo search algorithm. *Energy Conversion and Management*, *101*, 126–135.

Wang, G., Guo, L., Duan, H., Liu, L., Wang, H., Wang, B., et al. (2012). A hybrid meta-heuristic de/cs algorithm for ucav path planning. *Journal of Information and Computational Science*, *5*(16), 4811–4818.

Wróblewski, J. (1996). Theoretical foundations of order-based genetic algorithms. *Fundamenta Informaticae*, *28*(3, 4), 423–430.

Xiang-Tao, L., & Ming-Hao, Y. (2012). Parameter estimation for chaotic systems using the cuckoo search algorithm with an orthogonal learning method. *Chinese Physics B*, *21*(5), 050507.

Xu, H. J., Liu, J. K., & Lu, Z. R. (2016). Structural damage identification based on cuckoo search algorithm. *Advances in Structural Engineering*, p. 1369433216630128.

Yang, X.-S. & Nature-Inspired Metaheuristic Algorithms. (2008). *Luniver press*. UK: Beckington.

Yang, X.-S. (2010a). Firefly algorithm. *Engineering Optimization*, pp. 221–230.

Yang, X.-S. (2010b). A new metaheuristic bat-inspired algorithm. In *Nature inspired cooperative strategies for optimization (NICSO 2010)*, pp. 65–74. Springer.

Yang, X.-S., & Deb, S. (2009). Cuckoo search via lévy flights. In *Nature & Biologically Inspired Computing, 2009. NaBIC 2009. World Congress on*, pp. 210–214. IEEE.

Yang, X.-S., & Luniver Press. (2010). Nature-inspired metaheuristic algorithms second edition.

Yang, X.-S., & Deb, S. (2010). Engineering optimisation by cuckoo search. *International Journal of Mathematical Modelling and Numerical Optimisation*, *1*(4), 330–343.

Yang, X.-S., & Deb, S. (2014). Cuckoo search: Recent advances and applications. *Neural Computing and Applications*, *24*(1), 169–174.

Yasar, M. (2016). Optimization of reservoir operation using cuckoo search algorithm: Example of adiguzel dam, denizli, turkey. *Mathematical Problems in Engineering*.

Yildiz, A. R. (2013). Cuckoo search algorithm for the selection of optimal machining parameters in milling operations. *The International Journal of Advanced Manufacturing Technology*, *64*(1–4), 55–61.

Zhan, Z.-H., & Zhang, J. (2009). Discrete particle swarm optimization for multiple destination routing problems. In *Workshops on Applications of Evolutionary Computation*, pp. 117–122. Springer.

Zhao, H., Jiang, Y., Wang, T., Cui, W., & Li, X. (2016). A method based on the adaptive cuckoo search algorithm for endmember extraction from hyperspectral remote sensing images. *Remote Sensing Letters*, *7*(3), 289–297.

Zheng, H., & Zhou, Y. (2012). A novel cuckoo search optimization algorithm based on gauss distribution. *Journal of Computational Information Systems*, *8*(10), 4193–4200.

Zhou, Y., & Zheng, H. (2013). A novel complex valued cuckoo search algorithm. *The Scientific World Journal*.

Zhou, Y., Zheng, H., Luo, Q., & Jinzhao, W. (2013). An improved cuckoo search algorithm for solving planar graph coloring problem. *Applied Mathematics & Information Sciences*, *7*(2), 785–792.

Chapter 4
Methodology of Extracting the ODF Maxima Using CSA

4.1 Introduction

This chapter discusses and summarizes the research methodology that has been provided to achieve the objectives of the present research. This methodology includes three main stages. First, a cuckoo search algorithm is adapted for orientation distribution function. Second, three enhanced versions of the adaptive cuckoo search algorithm are presented to overcome the weaknesses of the original algorithm. Third, the best version of the algorithm is applied to real data on the human brain. The three stages provide five main contributions to the body of knowledge, thereby achieving the research objectives.

4.2 Schema of the Methodology

Figure 4.1 represents the processes to obtain the ODF and how will dealing with it in this research, generally (i.e., a brief overview of the mechanism of this research). The concise execution plan is divided into three main stages as depicted in Fig. 4.2. The first stage includes data preparation and adaptation. The second stage involves the enhancement of adaptation. The third stage uses real data on the human brain to evaluate the efficiency of the best algorithm version among CSA-ODF, MCSA-ODF, CSBA-ODF, and CSAHC-ODF

M. Shehab, *Artificial Intelligence in Diffusion MRI*, Studies in Computational Intelligence 877, https://doi.org/10.1007/978-3-030-36083-2_4

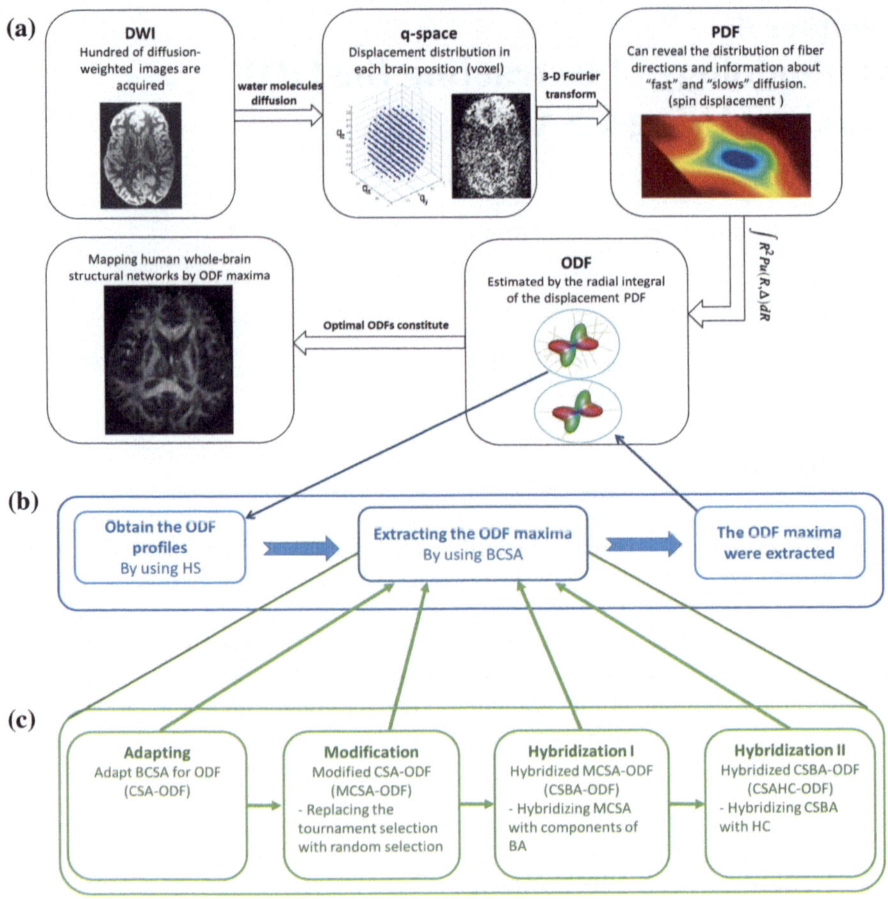

Fig. 4.1 Flowchart for dissertation

4.3 Stage1: Data Processing and Adaptation

4.3.1 Preparing Data

Two types of data are prepared in this section. First, synthetic data are generated to validate the mathematical models by comparing the behavior of real data with that of the data generated through the model. Subsequently, real data are acquired from volunteers to ensure the efficiency and accuracy of the work.

1. **Synthetic data**:
   Synthetic data are generated using the multitensor model. This model, which was also used in (Vitali et al. 2002; Hess et al. 2006; Tuch 2004; Alexander and Barker 2005; Shehab et al. 2016), assumes that each fiber has a Gaussian distribution that

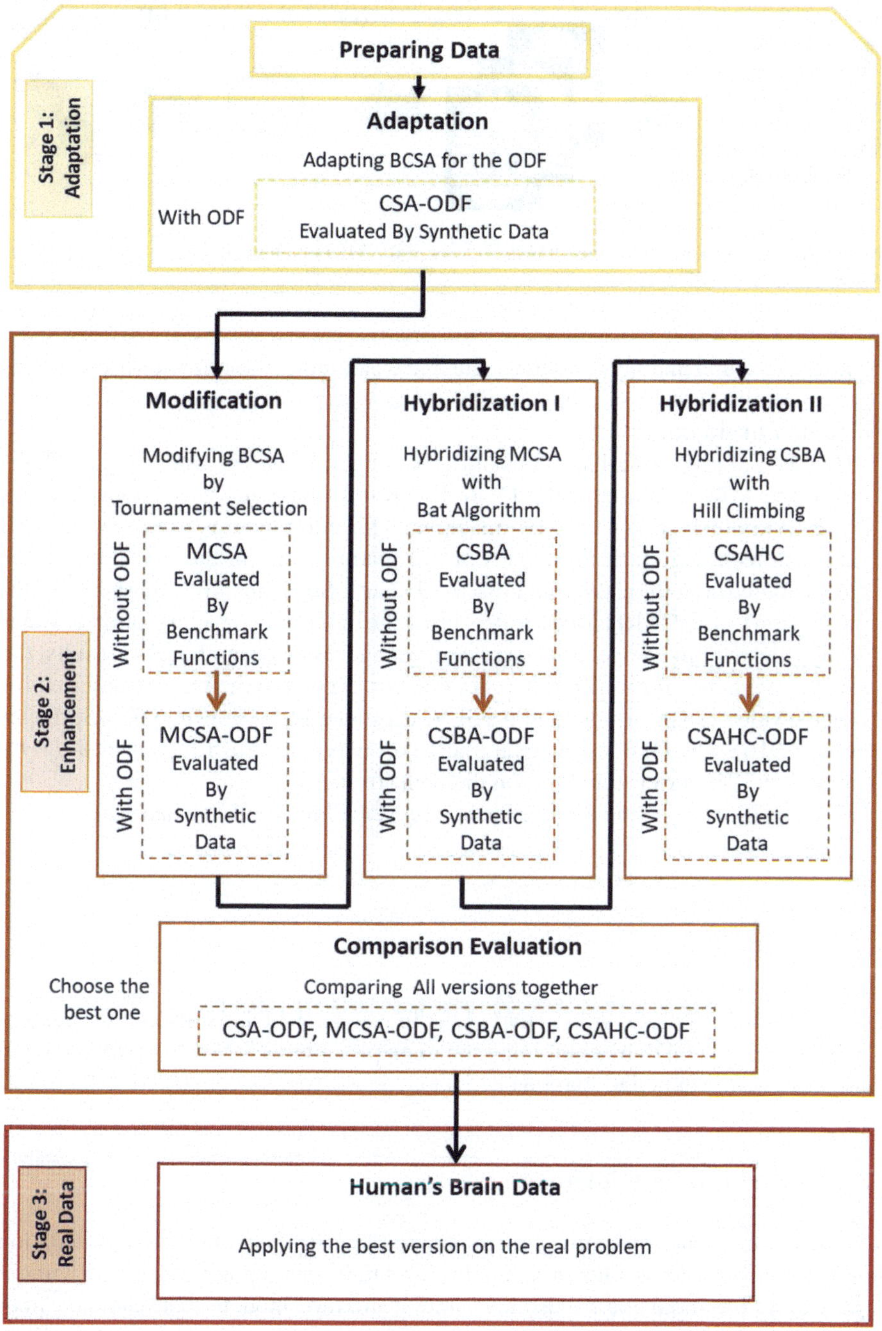

Fig. 4.2 Schema of the methodology

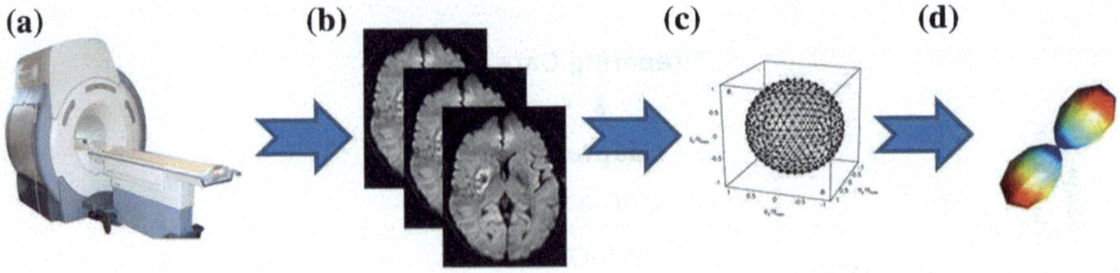

Fig. 4.3 The steps to preparing the ODF as follows: **a** the MRI machine, **b** the DWI, **c** the q-space, **d** ODF

describes it and that fibers do not exchange water molecules. Section 5.2 provides more details of the generating synthetic data which used in this research.

2. **Human brain data**:
This section presents all the steps of the water displacement function (i.e., ODF) from the MRI machine to the ODF. Figure 4.3 illustrates these steps, starting with (A), which shows the MRI machine. This equipment is used in a medical imaging technique to form images of the anatomy and physiological processes of the body in both healthy and diseased states. Subsequently, (B) shows DWI, which is a form of MRI based on measuring the random Brownian motion of water molecules within a voxel of tissue (Douis et al. 2016). From the DWI results, the signal intensities for a specifics voxel that correspond to the images acquired for each q value can be arranged into a 3D Cartesian grid, as shown in (C) (Sprenger et al. 2016). The ODF estimate at that voxel is provided as illustrated in (D) by applying 3D Fourier transform on the data (from C).
Notably, the data acquisition process is presented in detail in Chap. 8.

4.3.2 *Adaptation*

This section describes the second part of the first stage, which focuses on the adaptation BCSA for the ODF by combining the main procedures for BCSA with the ODF contents to detect the ODF maxima.

4.3.2.1 Problem Description

A mathematical function is always available to describe the (continuous) orientation density; this function is known as ODF. The ODF maxima are computed without any assumption about their existence, unlike mixture models that calculate fiber directions by describing the diffusion signal as the sum of finite discrete unidirectional components (Aganj et al. 2010). Therefore, extracting the maxima of ODF is not a

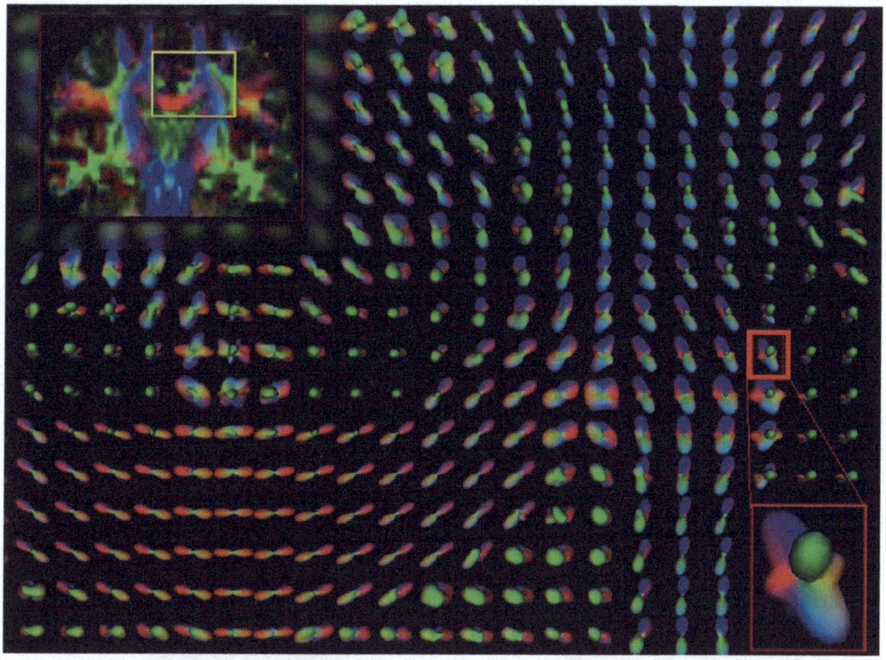

Fig. 4.4 A zoom on a single ODF (red box in the lower right) shows the multiple fiber orientations captured

trivial issue. Figure 4.4 shows the whole brain at the top left and a zoomed-in view of an ODF at the bottom right.

ODF is a spherical function defined by two spherical coordinates: θ and ϕ. In this case, our search space is spherical with $SH = (x_1, x_2, ..., x_u) = (\theta, \phi)$, and the objective function is merely the magnitude of an ODF that is used to evaluate the fitness of the solutions. In most cases, an ODF can have more than one maximum. These maxima correspond to the multiple fibers crossing in the considered voxel, and detecting all of them is important. Exclusion areas must be created around the detected maximum to perform the algorithm search elsewhere instead of within the vicinity of the already detected maximum. The number of exclusion areas varies dynamically with time and with the progress of the maximum detection process. Another important issue is the fact that ODF is an antipodally[1] (shown in Fig. 4.5) symmetric spherical function (Ghosh et al. 2008). In this case, creating an exclusion area around the antipodally symmetric coordinates of the detected maximum is important. A threshold is set arbitrarily at half the magnitude of the first detected

[1]The antipodal point is a point on the surface of a sphere is the point which is diametrically opposite to it. That situated that a line drawn from the one to the other passes through the center of the sphere and forms a true diameter.

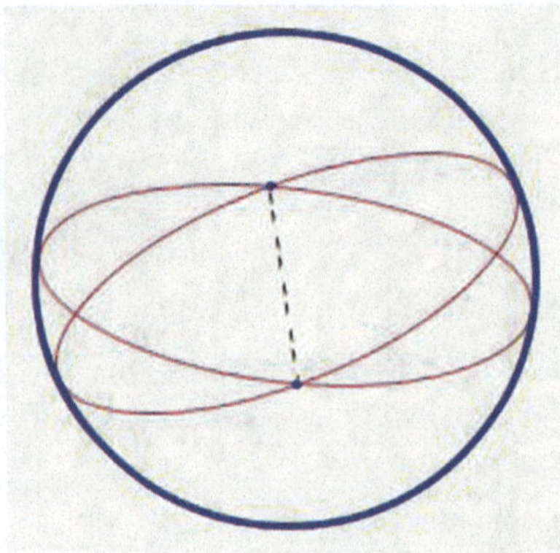

Fig. 4.5 Antipodal point

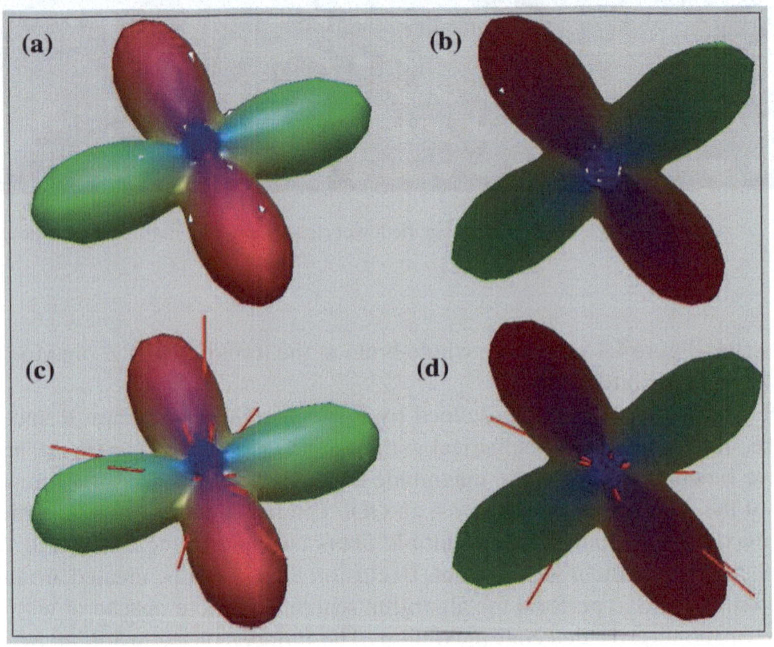

Fig. 4.6 **a**, **b** the white spots are the positions of the direction on the two faces of the ODF, **c**, **d** the red lines are their corresponding directions (maxima)

maximum to avoid the detection of small maxima. Figure 4.6 shows the detected maxima.

- **Spherical Harmonic Basis**
  The spherical harmonic expansion of a function represents the generalization of the Fourier transform to the domain of the sphere, thereby motivating its use for the

frequency domain representation of spherical functions in various scientific applications (Khalid et al. 2014). The concept is as follows: a signal on a sphere can be represented in terms of standard polar angles (θ, ϕ) or coordinates in a rotated frame. When the Wigner rotation matrix is used, a simple relation is established between SHs evaluated at (θ, ϕ) and SHs evaluated at the same physical angle but expressed in the rotated frame. This relation is used to evaluate and simplify FRT by expressing the integral in a carefully selected rotated frame.

For example, the representation of an ODF using the spherical harmonic basis is an established method in materials science, such as chemical engineering (Ritchie and Kemp 1999), energy (Simmonett et al. 2014), and data mining (Keller 2015). Moreover, a considerable body of literature in this field has described its use in characterizing crystal structures (Bunge 2013). Notably, the harmonic series has also been applied to the HARDI reconstruction problem via spherical deconvolution (Tournier et al. 2004) and to the analysis of multimodal diffusion using apparent diffusion coefficient (ADC) profiles (Alexander et al. 2002). Descoteaux et al. (2007) proposed an approach to estimate the coefficients of such decomposition as follows:

$$S(\theta_i, \phi_i) = \sum_{j=1}^{R} c_j Y_j(\theta_i, \phi_i) \tag{4.1}$$

where $R = (l + 1)(l + 2)/2$ is the number of terms in the modified SH basis Y of order l, c_j is SH coefficient and Y_j is the spherical harmonics components, defined as:

$$Y_l^m(\theta, \phi) = \sqrt{\frac{2l + 1}{4\pi} \frac{(l - m)!}{(l + m)!}} P_l^m(cos\theta)e^{-im\phi} \tag{4.2}$$

where m is a degree, θ and ϕ. The modified versions of the spherical basis functions are used to consider the fact that as a spherical function, the ODF is real and antipodally symmetric. This approach, which was proposed by Descoteaux et al. (2007), uses real symmetric spherical harmonics (RSSH) as follows:

$$Y_j = \begin{cases} \sqrt{2}Re(Y_l^{|m|}), & if\ m < 0 \\ Y_l^{|m|}, & if\ m = 0 \\ \sqrt{2}(-1)^{m+1}Im(Y_l^m), & if\ m > 0 \end{cases} \tag{4.3}$$

$$j(l, m) = (l2 + l + 2)/2 + m \tag{4.4}$$

where $Re(Y_l^{|m|})$ and $Im(Y_l^m)$ represent the real imaginary parts of Y_l^m respectively.

- **Objective Function**

  The objective function for evaluating a spherical harmonic solution x for an ODF can be formally defined based on the notations provided in Table 4.1. Figure 4.7 shows the steps in determining the objective function. The process starts from the

Table 4.1 The CSA terms in the ODF

ODF terms		CSA terms
Fibers	↔	Nest
SH	↔	Egg
Coordinates	↔	Objective function

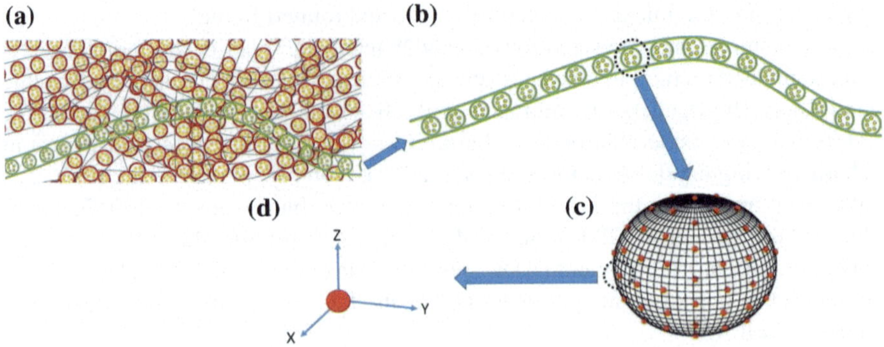

Fig. 4.7 **a** represents the search space (fibers), **b** represents one fiber (nest), **c** represents a SH (egg), **d** represents the coordinates each point in the SH

search space of numerous fibers. Each fiber contains many SHs, each SH contains many points, and each point is represented by coordinates x, y, and z.

The objective function *f(x)* is given in Eq. 4.5. The value of *f(x)* represent the spherical coordinates (θ, ϕ). Therefore, the objective function in this case refers to the magnitude of the ODF.

$$Max\ S(\theta_i, \phi_i) = \sum_{j=1}^{R} c_j Y_j(\theta_i, \phi_i) \tag{4.5}$$

4.4 Stage2: Enhancement Variations of Cuckoo Search Algorithm

This section presents the proposed enhancements for BCSA, which are grouped into three versions. The first version is modified BCSA (MCSA), in which tournament selection is used instead of random selection (basic selection) (Shehab et al. 2017; Shehab and Khader 2018). After ensuring its efficiency by evaluating it on a set of benchmark functions and comparing it with other algorithms, MCSA is used for an ODF, namely, MCSA-ODF. MCSA is evaluated using synthetic data. The second version, named CSBA, involves the hybridization of MCSA with the components of BA. The evaluation starts with a set of benchmark functions and comparisons with other algorithms, including BCSA and MCSA. CSBA is then used for an ODF, namely, CSBA-ODF. The third version, called CSAHC, merges BCSA with HC. Similar to the previous versions of the evaluations, CSAHC-ODF is symbolized by using CSAHC for an ODF.

Each version of CSA-ODF is compared with the previous one (i.e., MCSA-ODF with CSA-ODF, CSBA-ODF with MCSA-ODF, and so on). The summary of comparisons is provided at the end of all the versions. The results of all the versions are collected in the same section to facilitate comparison and to select the best version. Additional details of these versions and evaluation processes are discussed in the succeeding sections.

4.4.1 Modification BCSA

This section presents MCSA and its parameter tuning. In BCSA, the new solution is selected randomly for the first time. In this manner, the probability that a better solution will be chosen is minimal. Accordingly, the quality of the solution is affected, which leads to slow convergence (Shehab et al. 2017). Therefore, MCSA will be used by replacing random selection (current) with the tournament selection scheme, which works effectively in a wide range of problems. This technique can be implemented efficiently and is amenable to parallelization, thereby increasing diversity. In addition, the probability of obtaining better results is increased given the possibility of selecting the best solution (Shehab et al. 2018, 2019). With regard to parameter tuning, BCSA has two parameters, namely, number of nests (n) and probability of discovery (P_α). Parameter tuning is important to determine the solution and achieve the most ideal performance of MCSA. Therefore, the experiments use a set of values for each parameter as follows: $n = (5, 10, 15, 20, 50, 100, 150, 250, \text{and } 500)$ and $P_\alpha = (0.0, 0.1, 0.2, 0.3, 0.4, 0.5, 0.6, 0.7, 0.8, 0.9, \text{and } 1)$. One parameter will be changed at a time, but the other will remain constant to determine the optimum value of each parameter. MCSA is then applied to an ODF (MCSA-ODF). After testing its performance on a set of benchmark functions, MCSA is compared with other algorithms.

4.4.2 Hybridization Cuckoo Search Algorithm

The concept of hybridization is being increasingly developed in the field of computational intelligence. Hybridization mainly aims to enhance the weaknesses of an algorithm by hybridizing it with other algorithms or techniques. This section presents two types of hybridization, namely, hybridization with population-based algorithms by using the features of BA and hybridization with local search-based algorithms by adopting the advantages of HC.

4.4.2.1 Hybridization of MCSA with BA

MCSA is similar to any metaheuristic algorithm that aims to find the optimal solution. However, MCSA stores the best solution (i.e., the global solution) in each iteration without using it in the next iteration, thereby leading to a weak local searching capability and a slow convergence rate (Shehab et al. 2017). Therefore, BCSA is proposed to overcome the weaknesses of MCSA by hybridizing MCSA with the elements of BA. The main objective of such hybridization is to take advantage of BA features that directly apply the global best solution for the population to locate new positions for the solutions in each iteration. The best individuals among the entire population are used to enhance the convergence rate of MCSA. After validating its performance on a set of benchmark functions, CSBA is applied to the ODF problem, i.e., CSBA-ODF.

4.4.2.2 Hybridization of CSBA with HC

HC heuristic search is a mathematical optimization technique that belongs to the family of local search algorithms (Shehab et al. 2018, 2019). It is an iterative algorithm that starts with an arbitrary solution for a problem and then subsequently attempts to determine a better solution by incrementally changing one element of the solution. When the change produces a better solution, an incremental change is performed on the new solution (Abualigah et al. 2017). This process is repeated until no further improvement can be found. Therefore, we propose HC heuristic search by hybridizing it with CSBA, i.e., CSAHC, to refine the search space region of the new solution and identify the local optimal solution. This hybridization technique is performed to improve the local utilization of CSBA-ODF, and thus, achieve good-quality extraction of the ODF maxima. This application is called CSAHC-ODF in this dissertation.

4.4.3 Experiments and Evaluation Procedure

Three levels of evaluation are performed in this dissertation. The first step is evaluating the BCSA versions by using a set of benchmark functions to guarantee the efficiency of each version. The second step is evaluating the CSA-ODF versions by using synthetic data to select the best version. The third step is applying the best version of CSA-ODF on human brain data. The subsequent sections explain each step in detail.

1. **Benchmark Functions**:
   In the context of optimization, a common means to validate a new technique is by comparing its performance (in terms of solution quality) and robustness (in terms of the effect of the algorithm setting on the quality of the final solu-

Table 4.2 Benchmark functions

Symbol	Function	Definition
F1	Ackley	$f(x) = 20 + e - 20.e^{-0.2\sqrt{\frac{1}{n}\sum_{i=1}^{n} x_i^2}} - e^{\frac{1}{n}\sum_{i=1}^{n} cos(2\pi x_i)}$
F2	Griewank	$f(x) = \sum_{i=1}^{n} \frac{x_i^2}{4000} - \prod_{i=1}^{n} cos\left(\frac{x_i}{\sqrt{i}}\right) + 1$
F3	Penalty #1	$f(x) = \frac{\pi}{30}\left\{10sin^2(\pi y_1) + \sum_{i=1}^{n-1}(y_i - 1)^2 \bullet \left[1 + 10sin^2(\pi y_{i+1})\right]\right.$ $\left. + (y_n - 1)^2\right\} \sum_{i+1}^{n} u(x_i, 10, 100, 4),\ y_i = 1 + 0.25(x_i + 1)$
F4	Penalty #2	$f(x) = 0.1\left\{sin^2(3\pi x_1) + \sum_{i=1}^{n-1}(x_i - 1)^2 \bullet [1 + sin^2(3\pi x_{i+1})]\right.$ $\left. + (x_n - 1)^2[1 + sin^2(2\pi x_n)]\right\} + \sum_{i=1}^{n} u(x_i, 5, 100, 4)$
F5	Quartic with noise	$f(x) = \sum_{i=1}^{n}(i.x_i^4 + U(0, 1))$
F6	Rastrigin	$f(x) = 10.n \sum_{i=1}^{n}(x_i^2 - 10.cos(2\pi x_i))$
F7	Rosenbrock	$f(x) = \sum_{i=1}^{n-1}[100(x_{i+1} - x_i^2)^2 + (x_i - 1)^2]$
F8	Schwefel 2.26	$f(x) = 418.9829 \times D - \sum_{i=1}^{D} x_i sin(\lvert x_i\rvert^{\frac{1}{2}})$
F9	Schwefel 1.2	$f(x) = \sum_{i=1}^{n}\left(\sum_{j=1}^{i} x_j\right)^2$
F10	Schwefel 2.22	$f(x) = \sum_{i=1}^{n} \lvert x_i\rvert + \prod_{i=1}^{n} \lvert x_i\rvert$
F11	Schwefel 2.21	$f(x) = max_i\lvert x_i\rvert,\ 1 < i < n$
F12	Sphere	$f(x) = \sum_{i=1}^{n} x_i^2$
F13	Step	$f(x) = 6.n + \sum_{i=1}^{n} \lfloor x_i \rfloor$

tion) with those of other existing techniques on a set of benchmark optimization problems. Table 4.2 shows the details of 13 benchmark functions used in this dissertation. These functions are frequently used in the literature and presented in the Congress on Evolutionary Computation (CEC 2015) (Chen et al. 2014); the remaining functions are frequently used in the literature (Wang et al. 2015, 2016; Kabir et al. 2014).

The aforementioned functions are classified into two main groups: (i) unimodal optimization functions with only one local optimum and (ii) multimodal optimization functions that frequently contain multiple global and local optima

In this dissertation, the proposed versions of BCSA (i.e., MCSA, CSBA, and CSAHC) are investigated by using different types of optimization with various landscapes. The obtained results are compared with the following to validate the effectiveness of each version: (Gandomi and Alavi 2012), HS (Geem et al. 2001),

GA (Goldberg and Holland 1988), BA (Yang 2010), and the BCSA (Yang and Deb 2009). The results of the experiments show the best normalization (i.e., the best value for each method is determined and then normalized) and the mean normalization (i.e., the average value for each method is determined and then normalized) of each benchmark function for each algorithm. That is, normalization is the process of regularizing data with respect to the difference in values between samples. In the experiments, different benchmark functions are compared with one another. This procedure is difficult due to the wide gap between solutions. Therefore, normalization improves data integrity (Volkovs et al. 2014). In this research, normalization is calculated based on the following equation (Abdi 2009):

$$z_i = \frac{x_i - \mu}{S} \tag{4.6}$$

where is $x = (x_1, \ldots, x_n)$, n denotes the total number of data, z_i denotes the normalized data for element ith, μ is the mean and S is the standard deviation. Finally, the minimum element of the data will be 1 in the normalization results.

2. **Synthetic Data**:
As mentioned in Sect. 4.3.1, synthetic data are generated to evaluate each version of CSA-ODF. The following points describe several methods to evaluate these versions. These methods have been used in the literature (Prastawa et al. 2005; Pullens et al. 2010; McHugh et al. 2015; Yang et al. 2016).

    (a) *Angular Resolution*:
    Angular resolution related to the b-value. For example, when the b-value is high, the angular resolution quality is also high (i.e., directly proportional). However, increasing the b-value is typically realized by using long mixing times (Le Bihan et al. 2006). This procedure also minimizes the sensitivity of diffusion data to achieve a rapid diffusion effect (Pasternak et al. 2009). Therefore, another means to control angular resolution should be identified.
    This section presents the separation angles that can replace the b-value. That is, the separation angle is the angle between fibers that is affected by the sharp fiber ODF. For example, when the separation angle is small, the resolution will be high (i.e., inversely proportional). Thus, each version of CSA-ODF will be evaluated by determining the minimum separation angle between the two extracted maxima of the ODF. Notably, the angle starts from 90° to 0°.

    (b) *Influence of the parameters of BCSA*:
    Effects of BCSA parameters: (i) number of initial solutions and (ii) step size of *Lévy* flight. Each parameter has its own characteristics; however, both parameters exert the same effect on the search process, i.e., an increase in their values will increase processing time. The main objectives of using different values of the initial solutions and step size are as follows:
    - To determine the processing time for each version of CSA-ODF, namely, minimum time and maximum time.

- To evaluate the robustness of CSA-ODF versions by studying the effects of the parameters on their performance.
- To prove the optimal value of each parameter.

The initial solution and step size of *Lévy* flight have different values used in the experiments, i.e., (10, 20, 30, 40, 50, 60, 70, 80, 90, and 100) and (0.01, 0.05, 0.1, 0.5, 1, 1.5, 2, 2.5, 3, 3.5, and 4), respectively. The results of the initial solution will be compared with the fitted linear curve. The purpose of the fitted linear curve is to find slope and intercept values that define the line that is closest to the data. In other words, in applied mathematics, the numerical solution to solve the ordinary differential equation with the known exact solution (i.e., fitted linear curve). While the results of the of the step size will be determined the optimal value of the step size which refers to the minimum process time achieved to determine the best solution. The results of the comparisons among all the versions are presented in the summary of comparisons (Sect. 7.5).

(c) *Effect of the SNR and the b-value*

MRI can resolve the crossing fiber by acquiring additional information through sensitized scans (i.e., QBI) (Landman et al. 2012). However, the long echo time results in high noise, thereby reducing tissue contrast, which decreases the accuracy of QBI (Assemlal et al. 2008). With regard to the *b*-value, this value can affect the results by controlling the accuracy of the resolution as mentioned in part (a) Angular Resolution. In this research, the effect of the *b*-value is used to evaluate the performance of CSA-ODF.

This section presents the evaluation of the capability of each CSA-ODF version to withstand noise by measuring their performance with the effects of signal-to-noise ratio (SNR) and the *b*-value. The settings used in the experiments as follows: estimation order $l = 4$; SNR = (5, 10, 15, 20, 25, 30, 35, 40, 45, and 50); and the *b*-value = (1000, 1500, 2000, 2500, 3000, and 3500) (Hansen 1999; Sakaie and Lowe 2007). Finally, the results of each version will be compared with the results of BA, GA, KH, HS, and the previous version.

4.5 Stage2: Human Brain Data

The real data is described in Sect. 4.3.1. In this dissertation, CSA-ODF, MCSA-ODF, CSBA-ODF, and CSAHC-ODF will be compared (results of the synthetic data). Subsequently, the version that achieved best results will be applied to the real data set. Chapter 8 describes the details of data acquisition from Institut Perubatan dan Pergigian Termaju (IPPT), Universiti Sains Malaysia (USM), the implementation by using MATLAB, and the analysis through using FMRIB Software Library v5.0 (FSL).

4.6 Conclusion

This chapter summarizes the research methodology proposed in this dissertation. Four versions of the cuckoo search algorithm have been implemented for ODF.

The stages of our research are summarized in Fig. 4.2. The three proposed versions of BCSA are introduced sequentially. Each version focuses on specific weaknesses of the previous algorithm and improves them in different ways, thereby achieving high-quality solutions. Experiments are conducted on a set of benchmark functions to test each version of BCSA using a synthetic dataset to select the best version of CSA-ODF. The version that achieves the best results compared with those of the others will be applied to the human brain data. These procedures are described in detail in the succeeding chapters.

References

Abdi, H. (2009). *Experimental design and analysis for psychology*. Oxford University Press.

Abualigah, L. M., Sawaie, A. M., Khader, A. T., Rashaideh, H., Al-Betar, M. A., & Shehab, M. (2017). β-hill climbing technique for the text document clustering. *New Trends in Information Technology*, 60.

Aganj, I., Lenglet, C., & Sapiro, G. (2010). ODF maxima extraction in spherical harmonic representation via analytical search space reduction. In *International Conference on Medical Image Computing and Computer-Assisted Intervention* (pp. 84–91). Springer.

Alexander, D. C., & Barker, G. J. (2005). Optimal imaging parameters for fiber-orientation estimation in diffusion MRI. *Neuroimage, 27*(2), 357–367.

Alexander, D. C., Barker, G. J., & Arridge, S. R. (2002). Detection and modeling of non-gaussian apparent diffusion coefficient profiles in human brain data. *Magnetic Resonance in Medicine, 48*(2), 331–340.

Assemlal, H.-E., Tschumperlé, D., & Brun, L. (2008). Robust variational estimation of pdf functions from diffusion MR signal. In *Computational diffusion medical resonance imaging* (pp. 73–84).

Bunge, H.-J. (2013). *Texture analysis in materials science: mathematical methods*. Elsevier.

Chen, Q., Liu, B., Zhang, Q., Liang, J. J., Suganthan, P. N., & Qu, B. Y. (2014). *Problem definition and evaluation criteria for CEC 2015 special session and competition on bound constrained single-objective computationally expensive numerical optimization*. Technical report: Computational Intelligence Laboratory, Zhengzhou University, China and Nanyang Technological University, Singapore.

Descoteaux, M., Angelino, E., Fitzgibbons, S., & Deriche, R. (2007). Regularized, fast, and robust analytical q-ball imaging. *Magnetic Resonance in Medicine, 58*(3), 497–510.

Douis, H., Davies, M. A., & Sian, P. (2016). The role of diffusion-weighted MRI (DWI) in the differentiation of benign from malignant skeletal lesions of the pelvis. *European Journal of Radiology, 85*(12), 2262–2268.

Gandomi, A. H., & Alavi, A. H. (2012). Krill herd: a new bio-inspired optimization algorithm. *Communications in Nonlinear Science and Numerical Simulation, 17*(12), 4831–4845.

Geem, Z. W., Kim, J. H., & Loganathan, G. V. (2001). A new heuristic optimization algorithm: harmony search. *Simulation, 76*(2), 60–68 (2001).

Ghosh, A., Tsigaridas, E. P., Descoteaux, M., Comon, P., Mourrain, B., & Deriche, R. (2008) A polynomial based approach to extract the maxima of an antipodally symmetric spherical function and its application to extract fiber directions from the orientation distribution function in diffusion

MRI. In *11th International Conference on Medical Image Computing and Computer Assisted Intervention (MICCAI)* (p. 237).
Goldberg, D. E., & Holland, J. H. (1988). Genetic algorithms and machine learning. *Machine Learning, 3*(2), 95–99.
Hansen, P. C. (1999). *The L-curve and its use in the numerical treatment of inverse problems*. IMM, Department of Mathematical Modelling, Technical Universityof Denmark.
Hess, S. T., Girirajan, T. P., & Mason, M. D. (2006). Ultra-high resolution imaging by fluorescence photoactivation localization microscopy. *Biophysical Journal, 91*(11), 4258–4272.
Kabir, Md. W. U., Sakib, N., Chowdhury, S. M. R., Alam, M. S. (2014). A novel adaptive bat algorithm to control explorations and exploitations for continuous optimization problems. *International Journal of Computer Applications, 94*(13).
Keller, W. (2015). Data mining in grace monthly solutions. In *EGU General Assembly Conference Abstracts* (Vol. 17, p. 15195).
Khalid, Z., Kennedy, R. A., & McEwen, J. D. (2014). An optimal-dimensionality sampling scheme on the sphere with fast spherical harmonic transforms. *IEEE Transactions on Signal Processing, 62*(17), 4597–4610.
Landman, B. A., Bogovic, J. A., Wan, H., ElShahaby, F. El. Z. Bazin, P.-L. & Prince, J. L. (2012). Resolution of crossing fibers with constrained compressed sensing using diffusion tensor MRI. *NeuroImage, 59*(3), 2175–2186.
Le Bihan, D., Poupon, C., Amadon, A., & Lethimonnier, F. (2006). Artifacts and pitfalls in diffusion MRI. *Journal of Magnetic Resonance Imaging, 24*(3), 478–488.
McHugh, D. J., Zhou, F., Cristinacce, P. L. H., Naish, J. H., & Parker, G. J. M. (2015). Ground truth for diffusion MRI in cancer: A model-based investigation of a novel tissue-mimetic material. In *International Conference on Information Processing in Medical Imaging* (pp. 179–190). Springer.
Pasternak, O., Sochen, N., Gur, Y., Intrator, N., & Assaf, Y. (2009). Free water elimination and mapping from diffusion MRI. *Magnetic Resonance in Medicine, 62*(3), 717–730.
Prastawa, M., Bullitt, E., & Gerig, G. (2005). Synthetic ground truth for validation of brain tumor mri segmentation. *Medical Image Computing and Computer-Assisted Intervention-MICCAI, 2005*, 26–33.
Pullens, P., Roebroeck, A., & Goebel, R. (2010). Ground truth hardware phantoms for validation of diffusion-weighted mri applications. *Journal of Magnetic Resonance Imaging, 32*(2), 482–488.
Ritchie, D. W., & Kemp, G. J. L. (1999). Fast computation, rotation, and comparison of low resolution spherical harmonic molecular surfaces. *Journal of Computational Chemistry, 20*(4), 383–395.
Sakaie, K. E., & Lowe, M. J. (2007). An objective method for regularization of fiber orientation distributions derived from diffusion-weighted MRI. *NeuroImage, 34*(1): 169–176.
Shehab, M., & Khader, A. T. (2018). Modified cuckoo search algorithm using a new selection scheme for unconstrained optimization problems. *14*, 1.
Shehab, M., Daoud, M. Sh., AlMimi, H. M., Abualigah, L. M., & Khader, A. T. (2019). Hybridizing cuckoo search algorithm for extracting the ODF maxima in spherical harmonic representation. *International Journal of Bio-Inspired Computation*, (in press).
Shehab, M., Khader, A. T., & Al-Betar, M. A. (2016) New selection schemes for particle swarm optimization. *IEEJ Transactions on Electronics, Information and Systems, 136*(12), 1706–1711. https://doi.org/10.1541/ieejeiss.136.1706.
Shehab, M., Khader, A. T., & Al-Betar, M. A. (2017). A survey on applications and variants of the cuckoo search algorithm. *Applied Soft Computing*.
Shehab, M., Khader, A. T., & Alia, M. A. (2019). Enhancing cuckoo search algorithm by using reinforcement learning for constrained engineering optimization problems. In *2019 IEEE Jordan International Joint Conference on Electrical Engineering and Information Technology (JEEIT)* (pp 812–816). IEEE.
Shehab, M., Khader, A. T., & Laouchedi, M. (2017). Modified cuckoo search algorithm for solving global optimization problems. In *International Conference of Reliable Information and Communication Technology* (pp. 561–570). Springer.

Shehab, M., Khader, A. T., & Laouchedi, M. (2018). A hybrid method based on cuckoo search algorithm for global optimization problems. *Journal of ICT, 17*(3), 469–491.

Shehab, M., Khader, A. T., Al-Betar, M. A., & Abualigah, L. M. (2017). Hybridizing cuckoo search algorithm with hill climbing for numerical optimization problems. In *2017 8th International Conference on Information Technology (ICIT)* (pp. 36–43). IEEE.

Shehab, M., Khader, A. T., Laouchedi, M., & Alomari, O. A. (2018). Hybridizing cuckoo search algorithm with bat algorithm for global numerical optimization. *The Journal of Supercomputing*, 1–28.

Simmonett, A. C., Pickard, F. C. IV, Schaefer, H. F. III, & Brooks, B. R. (2014). An efficient algorithm for multipole energies and derivatives based on spherical harmonics and extensions to particle mesh ewald. *The Journal of Chemical Physics, 140*(18), 184101.

Sprenger, T., Sperl, J. I., Fernandez, B., Golkov, V., Eidner, I., Sämann, P. G., et al. (2016). Bias and precision analysis of diffusional kurtosis imaging for different acquisition schemes. *Magnetic Resonance in Medicine*.

Tournier, J.-D., Calamante, F., Gadian, D. G., & Connelly, A. (2004). Direct estimation of the fiber orientation density function from diffusion-weighted MRI data using spherical deconvolution. *NeuroImage, 23*(3), 1176–1185.

Tuch, D. S. (2004) Q-ball imaging *Magnetic Resonance in Medicine, 52*(6), 1358–1372.

Vitali, C., Bombardieri, B., Jonsson, R., Moutsopoulos, H. M., Alexander, E. L., Carsons, S. E., et al. (2002). Classification criteria for sjögren's syndrome: A revised version of the European criteria proposed by the American-European consensus group. *Annals of the Rheumatic Diseases, 61*(6), 554–558.

Volkovs, M., Chiang, F., Szlichta, J., & Miller, R. (2014). Continuous data cleaning. In *2014 IEEE 30th International Conference on Data Engineering (ICDE)* (pp. 244–255). IEEE.

Wang, G.-G., Gandomi, A. H., Yang, X.-S., & Alavi, A. H. (2016). A new hybrid method based on krill herd and cuckoo search for global optimisation tasks. *International Journal of Bio-Inspired Computation, 8* (5), 286–299.

Wang, L., Zhong, Y., & Yin, Y. (2015). A hybrid cooperative cuckoo search algorithm with particle swarm optimisation. *International Journal of Computing Science and Mathematics*, *6*(1), 18–29.

Yang, X.-S. (2010). A new metaheuristic bat-inspired algorithm. In *Nature inspired cooperative strategies for optimization (NICSO 2010)* (pp. 65–74). Springer.

Yang, X.-S., & Deb, S. (2009). Cuckoo search via lévy flights. In *World Congress on Nature & Biologically Inspired Computing, 2009. NaBIC 2009* (pp. 210–214). IEEE.

Yang, F., Ford, J., & Dogan, N. (2016). Su-f-r-57: Validation of quantitative radiomic texture features for oncologic mri: A simulation study. *Medical Physics, 43*(6), 3386–3386.

Chapter 5
Adaptive Cuckoo Search Algorithm for Extracting the ODF Maxima

5.1 Introduction

In this chapter, an initial investigation of adapting the basic cuckoo search algorithm (BCSA) for the orientation distribution function (ODF) is presented. This chapter aims to extract the maxima of the ODF using BCSA, namely, CSA-ODF.

This chapter comprises Four main sections. Section 5.2 illustrates the steps to generate the synthetic data. Followed by Sect. 5.3 describes the CSA-ODF. The experiments and the results, with a detailed analysis of the behavior of the proposed CSA-ODF in terms of convergence properties are represented in Sect. 5.4. Finally, Sect. 5.5 includes the conclusion of this chapter.

5.2 Preparing Data

This section shows the generating of the synthetic data which uses to evaluate the performance of CSA-ODF versions. The multitensor model is used to generate the synthetic data where it can be allowed to access of fiber different populations inside the voxel (Tuch et al. 2002; Alexander and Barker 2005; Descoteaux et al. 2006; Shehab et al. 2016). This model assumes that each fiber has a gaussian distribution that describes it and that fibers do not exchange water molecules. Equation 5.1 shows the expression for the signal in multitensor model.

$$E(q, \hat{q}) = E_0 \sum_{i=1}^{M} f_i exp(-b\hat{q}^{\tau} D^{(i)} \hat{q}) \tag{5.1}$$

where q and $\hat{q}$ are the length and unit vector, M is the number of fiber-compartments, b is the b-value (see Sect. 2.3.1 for more details), and τ is the diffusion-sensitizing gradient duration; the volume fraction of the ith compartment is denoted by f_i, so

M. Shehab, *Artificial Intelligence in Diffusion MRI*, Studies in Computational Intelligence 877, https://doi.org/10.1007/978-3-030-36083-2_5

that $\sum_{i=1}^{M} f_i = 1$; $D^{(i)}$ represents the second-order self-diffusion tensor of the ith compartment, and E_0 is the signal amplitude in the absence of magnetic field gradients (i.e., b = 0).

The ith self-diffusion tensor can be parameterized as $D^{(i)} = R^T(\theta, \phi)AR(\theta, \phi)$, where $R(\theta, \phi)$ is the rotation matrix or an operator that rotates a unit vector initially oriented along the x-axis toward the main axis (θ, ϕ) of the ith fiber. The diagonal matrix A contains information about the anisotropy of the diffusion process within each fiber compartment and can be written as:

$$A = \begin{bmatrix} \lambda_1 & 0 & 0 \\ 0 & \lambda_2 & 0 \\ 0 & 0 & \lambda_3 \end{bmatrix}$$

where λ_1 is the diffusivity along the main fiber-axis ($\lambda_1 \geq \lambda_2 \geq \lambda_3$), λ_2 and λ_3 are the diffusivities in the plane perpendicular to this axis.

Equation 5.2 illustrates the theoretical expression for the exact ODF corresponding to the diffusion tensor model

$$ODF_\tau(\hat{r}) = \frac{1}{Z^n}(\pi t)^{1/2} \sum_{i=1}^{M} f_i \cdot (\hat{r}^\tau D^{(i)-1} \hat{r})^{-1/2} \tag{5.2}$$

where Z^n is a normalization constant, and $D^{(i)-1}$ is the inverse of $D^{(i)}$. The following steps show the generating a set of 200 random fiber configurations with two fibers (the number of fiber populations detected in voxels with fiber crossings) (Shehab et al. 2017).

- set diffusion encoding gradient directions $g_i = (sin\theta_i cos\phi_i, sin\theta_i sin\phi_i, cos\theta_i)^t$ using certain order tessellation of the icosahedron with n samples on the sphere.
- set the number of fibers to a specific n fibers between 0 and 3 (0 is for isotropic signal).
- for each fiber $1 \leq k \leq n$, randomly choose an orientation (θ_k, ϕ_k) and a relative weight pk for the k-th fiber, where $\sum_{k=1}^{n} pk = 1$. In practice, we impose a minimum angle between fibers of 45° and relative weights between 0.3 and 0.7 for 2-fiber and between 0.2 and 0.4 for 3-fiber distributions. Otherwise, the fiber compartments are too weak to distinguish them.
- the profile with eigenvalues $[700, 700, 700] \times 10^{-6}$mm^2/s are used for the isotropic voxels. for the other distributions, we generate 3×3 tensors with eigenvalues $[\lambda_1, \lambda_2, \lambda_3] = [300, 300, 1700] \times 10^{-6}$mm^2/s and generate each fiber tensor d_k using rotation matrix rot to orient the main axis of the tensor in direction (θ_k, ϕ_k), i.e. $d_k = rot^{\tau}_{(\theta_k, \phi_k)} diag(\lambda_1, \lambda_2, \lambda_3) rot_{(\theta_k, \phi_k)}$. Note that we choose tensor profiles with anisotropy nearly as high as the most anisotropic regions found in the human brain. This facilitates the discrimination of 2- and 3-fiber distributions. The underlying assumption is that lower anisotropy levels are due to noise and partial volume averaging from fibers crossing, kissing or diverging (Tournier et al. 2004).

- pick a b-value and for each encoding direction i generate the diffusion signal

$$s(b, g_i) = s_0 \sum_{k=1}^{n} p_k e^{-b g_i^t d_k g_i} \tag{5.3}$$

- add complex gaussian noise with a standard deviation (std) of $\sigma = s_0/\zeta$ to the raw signal of Eq. 5.3. That is, to each $s(g_i)$, we add a random complex number with independent real and imaginary parts coming from a zero mean gaussian distributions with $\sigma = 1/\zeta$ ($s_0 = 1$) and take the modulus to obtain the noisy synthetic data. A typical value is $\zeta = 35$ so that the unweighted signal s_0 has a signal-to-noise ratio (snr) of 35.

5.3 Basic Cuckoo Search Algorithm for Extract the Maxima of the ODF

Compared with other techniques, QBI uses a simpler acquisition scheme with considerably fewer gradient directions, thereby making it attractive and practicable for clinical MRI systems. However, this technique exhibits several expensive mathematical complexities. Descoteaux et al. (2007) proposed the Funk-Hecke theorem and the spherical harmonic basis functions for ODF reconstruction. The extraction of the ODF maxima is an important step before reconstructing the fiber tracts (tractography) because fiber orientations are confounded by the angular information of the maxima. Numerous strategies have been developed for this purpose. Some of these strategies are analytical, whereas others are numerical, such as Ghosh (1977), Bloy and Verma (2008), Descoteaux et al. (2006), Aganj et al. (2010) and Laouchedi et al. (2014).

This section presents the procedures for extracting the ODF maxima via CSA. The proposed method is expected to extract all ODFs accurately and efficiently (Shehab et al. 2018; Abualigah et al. 2017). Moreover, numerous search strategies have been introduced to reduce processing time (Shehab et al. 2017). The next section provides the complete description of CSA-ODF, followed by the experiments on CSA-ODF using synthetic data.

5.3.1 Method

The principle behind BCSA is described as follows. Each cuckoo bird selects a random nest from the search space, i.e., Fiber, $\{F_d\}_{d=1,2,3,\ldots,m}$ and lays one egg, i.e., spherical harmonic (SH), $\{SH_i\}_{i=1,2,3,\ldots,n}$ where $SH_i = (x_1, x_2, \ldots, x_u)$ coordinates and $x_1 = (\theta_1, \phi_1)$. The nest, which contains high-quality eggs, will be moved to the next generation. The probability of survival of an alien egg in the host nests is P_α

[0, 1] (i.e., if an alien egg is discovered by the host bird, it will either be thrown out or the host bird will abandon the nest) (Shehab et al. 2017). Algorithm 2 presents the pseudo-code for the CSA-ODF algorithm, followed by the main steps of the optimization strategies.

Algorithm 2 CSA-ODF

Objective function $f(X)$
while $t < Max\_iterations$ **do**
  Generate initial population includes SH coordinates $xi_{(i=1,2,...,u)}$, where $x_i = (\theta_i, \phi_i)$
  while stopping criterion **do**
    Get a solution by *Lévy* flights: Max_i /** assume that the best solution**/
    Check if the Max_i in the search space
    Evaluate its quality/ fitness $F(Max_i)$ by Eq.4.5
    Get new two solutions randomly Max_j and Max_k /** neighbors of the Max_i **/
    Check if the solutions Max_j and Max_k are in the search space
    Evaluate its quality/ fitness $F(Max_j)$ and $F(Max_k)$ by Eq.4.5
    Get the Minimum and Maximum between Max_i, Max_j, and Max_k : *MIN* and *MAX* /** If the *MIN* pass from the last checking Pα the other will pass **/
    Check the *MIN* by Probability $(P\alpha)$ /** if the MIN discover: will be abandoned the solutions and build new solution by *Lévy* flights **/
  end while
  Maximum= Max
  Store the other two solutions (precaution)
  Labeling the points that constitute the vicinity of the detected *Maximum* and its antipodal
end while
Postprocess results and visualization

- **Step 1. Input**:
  Population, $P = (F_1, F_2, F_3, ..., F_n)$, n paths. Where $F_1 = (SH_1^1, SH_2^1, SH_3^1, ..., SH_m^1)$, m spherical harmonics, and SH_1^1 is the starting point of F_1 and SH_m^1 its endpoint. $SH_i = (x_1, x_2, x_2, ..., x_u)$ coordinates and $x_i = (\theta_i, \phi_i)$.
- **Step 2. Use *Lévy* flight behavior to determine the next point**:
  One of the most powerful features of BCSA is the use of *Lévy* flights to select new solutions (Shehab et al. 2018b). In this approach, the selected point x_{i+1} is produced by perturbing the current xi with a change of position $((L_i))$. A random step $((s_i))$ is generated by a symmetric *Lévy* distribution (Mantegna 1994). The following equation is used to produce s_i:

$$s_i = \frac{u}{|v|^{1/\beta}} \tag{5.4}$$

Parameters u and v in the preceding equation are expressed by the normal distributions in Eq. 5.5. The variance and sigma calculate by using Eq. 5.6 with $1 \leqslant \beta \leqslant 2$ and where Γ (.) is the gamma function.

$$u \sim N\left(0, \sigma_u^2\right), \qquad v \sim N\left(0, \sigma_v^2\right) \tag{5.5}$$

$$\sigma_u = \left\{ \frac{\Gamma(1+\beta)\, sin\,(\pi\beta/2)}{\Gamma\left[(1+\beta)/2\right]\beta 2^{(\beta-1)/2}} \right\}^{1/\beta}, \sigma_v = 1 \tag{5.6}$$

Once s_i has been calculated, the required change of position L_i is computed as follows:

$$L_i = 0.01.s_i \bigoplus (x_i - x_{best}) \tag{5.7}$$

where the product $\bigoplus$ denotes entrywise multiplications whereas x_{best} is the best solution found thus far. Finally, the new candidate point, x_{i+1}, is calculated by using

$$x_{i+1} = x_i + L_i \tag{5.8}$$

- **Step 3. Elitist Selection Strategy**:
  After producing x_{i+1}, it must be compared with its past value x_i. If the fitness value of x_{i+1} is better than x_i, then x_{i+1} is accepted as the final solution. Otherwise, x_i is retained (Shehab et al. 2019). This procedure can be presented by the following expression:

$$x_{i+1} = \begin{cases} \mathrm{x}_{i+1}, & \text{if } x_{i+1} > x_i, \quad \text{Comparing the fitness calculated by Eq.4.5} \\ \mathrm{x}_i, & otherwise \end{cases} \tag{5.9}$$

- **Step 4. Check with the probability** $P\alpha$:
  A uniform random number (r_1) is generated within the range [0, 1] to implement this operation. When (r_1) is less than P_α, the individual x_i is selected and modified according to Eq. 5.10. Otherwise, x_i remains unchanged (Shehab and Khader 2018). This operation is represented by the following model:

$$x_i = \begin{cases} \mathrm{x}_i + rand.\left(x_{k_1} - x_{k_2}\right), & \text{if}(r_1 < P_\alpha) \\ \mathrm{x}_i, & otherwise \end{cases} \tag{5.10}$$

  where *rand* is a random number from a normal distribution, whereas k_1 and k_2 are random integers from 1 to n.
- **Step 5. Check the stopping criteria**:
  At this point, we evaluate the ODF and its neighbors to determine whether a visited point $x_i = (\theta_i, \phi_i)$ is the maximum.

$$x_{\theta_{i+1},\phi_i} = (\theta_{i+1}, \phi_i) \tag{5.11}$$

$$x_{\theta_{i-1},\phi_i} = (\theta_{i-1}, \phi_i) \tag{5.12}$$

$$x_{\theta_i,\phi_{i+1}} = (\theta_i, \phi_{i+1}) \tag{5.13}$$

$$x_{\theta_i,\phi_{i-1}} = (\theta_i, \phi_{i-1}) \tag{5.14}$$

The stopping criterion is achieved when the maximum bending angle of the curve, which can happen by:

$$Max_\theta = (f(\theta_i, \phi_i) - f(\theta_{i+1}, \phi_i)).(f(\theta_i, \phi_i) - f(\theta_{i-1}, \phi_i)) \tag{5.15}$$

and

$$Max_\phi = (f(\theta_i, \phi_i) - f(\theta_i, \phi_{i+1})).(f(\theta_i, \phi_i) - f(\theta_i, \phi_{i-1})) \tag{5.16}$$

5.4 Experiments and Results

5.4.1 Experimental Design

A series of experiments was conducted to discuss the behavior of CSA-ODF. From an angular resolution, which was used to enhance the accuracy of fiber orientation, the experiments began with a separation angle of 90°, an initial solution of 20, and no noise. Then, the influences of the control parameters, which include initial solutions and *Lévy* flights, were considered. This step measured the effectiveness of CSA-ODF by using different values of the aforementioned parameters, i.e., initial solution = (10, 20, 30, 40, 50, 60, 70, 80, 90, and 100) and *Lévy* flights = (0.01, 0.4, 0.75, 1.2, 1.6, 2, 2.4, 2.8, 3.2, 3.6, and 4). The third part of the experiment focused on the robustness of CSA-ODF by using different values of SNR (5, 10, 15, 20, 25, 30, 35, 40, 45, and 50) and the *b*-value (1000, 1500, 2000, 2500, 3000, and 3500) when comparing its performance with those of BA, KH, GA, and HS. In the experiments, each *b*-value will be fixed for all SNR values.

5.4.2 Experimental Results and Discussion

5.4.2.1 Angular Resolution of the CSA-ODF

The angular high generally describes the capability of any image-forming device. QBI is a high-angular-resolution diffusion imaging (HARDI) method that is capable of resolving complex fiber structures (i.e., the application of the QBI model to a HARDI acquisition scheme allows the estimation of the ODF). However, the accuracy and angular resolution of QBI for defining fiber orientation and its dependence on diffusion imaging parameters remain unclear. Therefore, angular resolution is considered one of the important factors for evaluating CSA-ODF.

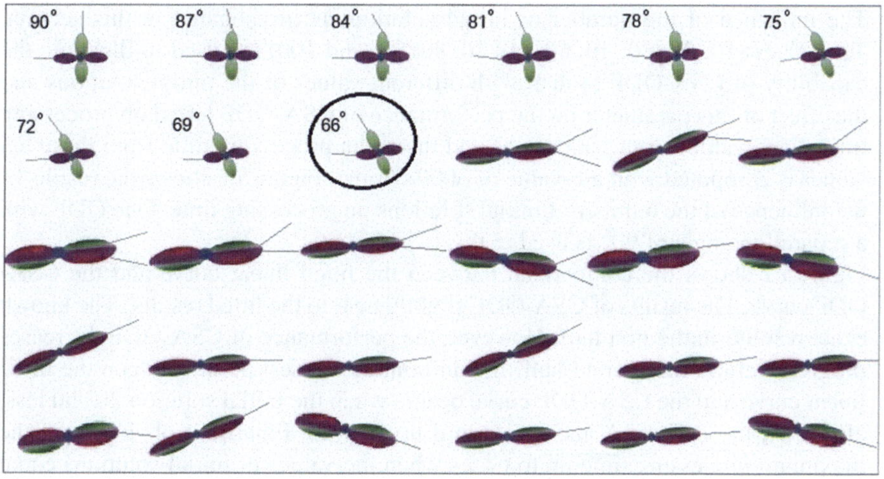

Fig. 5.1 The ability of the CSA-ODF to detect maxima separated by 66°

Noise-free synthetic data were obtained as described in Sect. 4.4.3 with 20 initial solutions and by adopting the optimization strategies described in Sect. 5.3.1. As demonstrated in many studies, QBI-ODF reconstructed using RSSH of the fourth order are relatively smooth and angular resolutions obtained using various algorithms are approximately 63° and 60°, such as Aganj et al. (2010) and Descoteaux et al. (2009), respectively. These resolutions are sufficient for the clinical use of tractography. Wedeen et al. (2005) showed that most of the intersections in the brain exist at approximately 90°. CSA-ODF demonstrated its capability to detect the ODF maxima separated by 66° and above, as shown in Fig. 5.1. The ODF at the upper left corner presents a separation angle of 90° between its maxima. This angle is decreased by 3° each time to produce the remaining ODFs. Figure 5.1 shows that two fibers are detected for separation angles of approximately 66°.

5.4.2.2 Influence of Control Parameter

This section discusses the effects of the CSA parameters on the performance of CSA-ODF. The results of the experiments presented in this section will be used to compare the stability of CSA-ODF with those of the other proposed methods (i.e., MCSA-ODF, CSBA-ODF, and CSAHC-ODF) under different values of the parameters. The effects of the two parameters of CSA are illustrated as follows.

- **Influence of the number of the initial solutions**
  The number of initial solutions is an important parameter in the search procedure. It is directly related to the manner in which the search space is covered, and it increases certainty in finding the maxima. However, the number of operations also increases in each iteration, which lengthens processing time (Shehab et al. 2019).

The influence of the number of initial solutions is investigated in this section. Ten values (10, 20, 30, 40, 50, 60, 70, 80, 90, and 100) are used to illustrate the capability of CSA-ODF to deal with different values of the initial solutions and the effect of this parameter on the performance of CSA-ODF based on processing time. Each value is run 1000 times and the mean processing time from these ten values is computed with a b-value of $3000\,\text{s/mm}^2$. Figure 5.2 shows the results of the influence of the number of initial solutions on processing time. One ODF with a separation angle of 90° is used in this experiment.
Figure 5.2 shows the comparison between the fitted linear curve and the CSA-ODF curve. The results of CSA-ODF slightly near to the fitted results (The known exact results) in the first half. However, the performance of CSA-ODF decreases progressively in the second half. In addition, the closest point between the fitted linear curve and the CSA-ODF curve occurs when the initial solution 20 and less. after that, the difference increases until up to 4.2 s. Finally, it obvious that the maximum processing time up to 64.2 s when the values of initial solutions equal 100 (i.e., The largest value), this indicates to the weakness of the CSA-ODF by dealing with large numbers of the initial solutions.

- **Influence of the *Lévy* Flights**
  The random walk step used in Eq. 3.1 is drawn from the *Lévy* distribution. This section presents a set of values for this step, which is used to span the search space. Eleven values (0.01, 0.05, 0.1, 0.5, 1, 1.5, 2, 2.5, 3, 3.5, and 4) are selected. Each value is run 1000 times and the mean of the eleven values is computed with the b-value= $3000\,\text{sec/mm}^2$.
  As stated in Sect. 5.3.1, *Lévy* flight is used in the second step of the adaptation with a set of different values. For example, when used in a step with a large value to avoid searching for a long time, *Lévy* flight may jump out of the search space without finding the near solution. By contrast, the use in a small step size value leads to slow convergence, which takes a long time. Therefore, the step size value should be determined carefully. In addition, it can notice that the bending becomes less when the value of step size equal 2 and above (i.e., The best solution achieved by the MCSA-ODF When the value of the step size is approximately equal to 2). Figure 5.3 shows that when the step size is changed, processing time is evidently affected. That is, the performance of CSA-ODF is weak because it does not have the capability to maintain its level in case step size increases. In other words, the CSA-ODF achieved the best solution when the value of step size equal 4 (i.e., The largest step size), thereby leading to difficulties, as mentioned earlier. In addition, the difference in processing time between the largest value and the smallest value of the step size is approximately 7.5 s. The time difference indicates that CSA-ODF is significantly influenced by step increments.

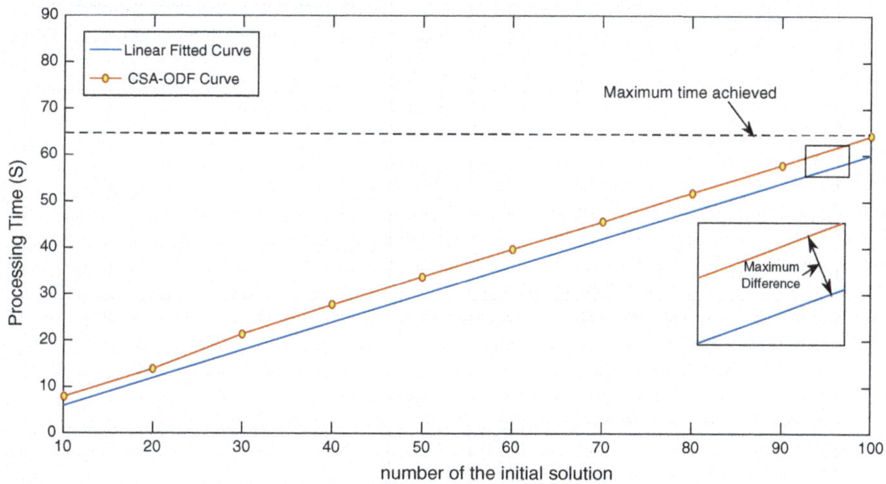

Fig. 5.2 Influence of the number of initial solutions on the processing time of the CSA-ODF

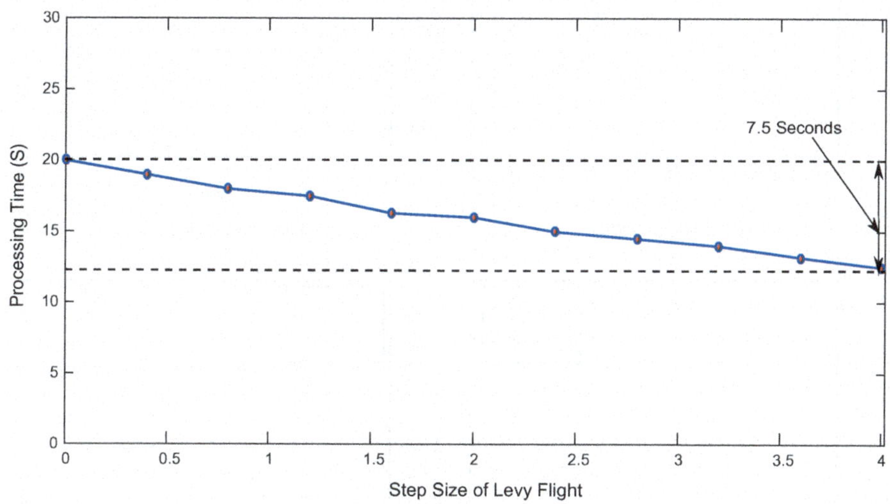

Fig. 5.3 Influence of the step size of *Lévy* Flights on the processing time of the CSA-ODF

5.4.2.3 Influence of SNR and *b*-value

DWI is known as a noisy modality (Martín et al. 2016), and noise is present not only in the acquired data but also in the reconstructed invariants, such as fractional anisotropy (FA) and mean diffusion (MD) (Scott et al. 2016). It is also present in the estimated tensors (Abbasloo et al. 2016) and ODFs (Veraart et al. 2016). Thus, noise occasionally leads to missing one or two maxima even with a separation angle of 90°, which affects ODF shape.

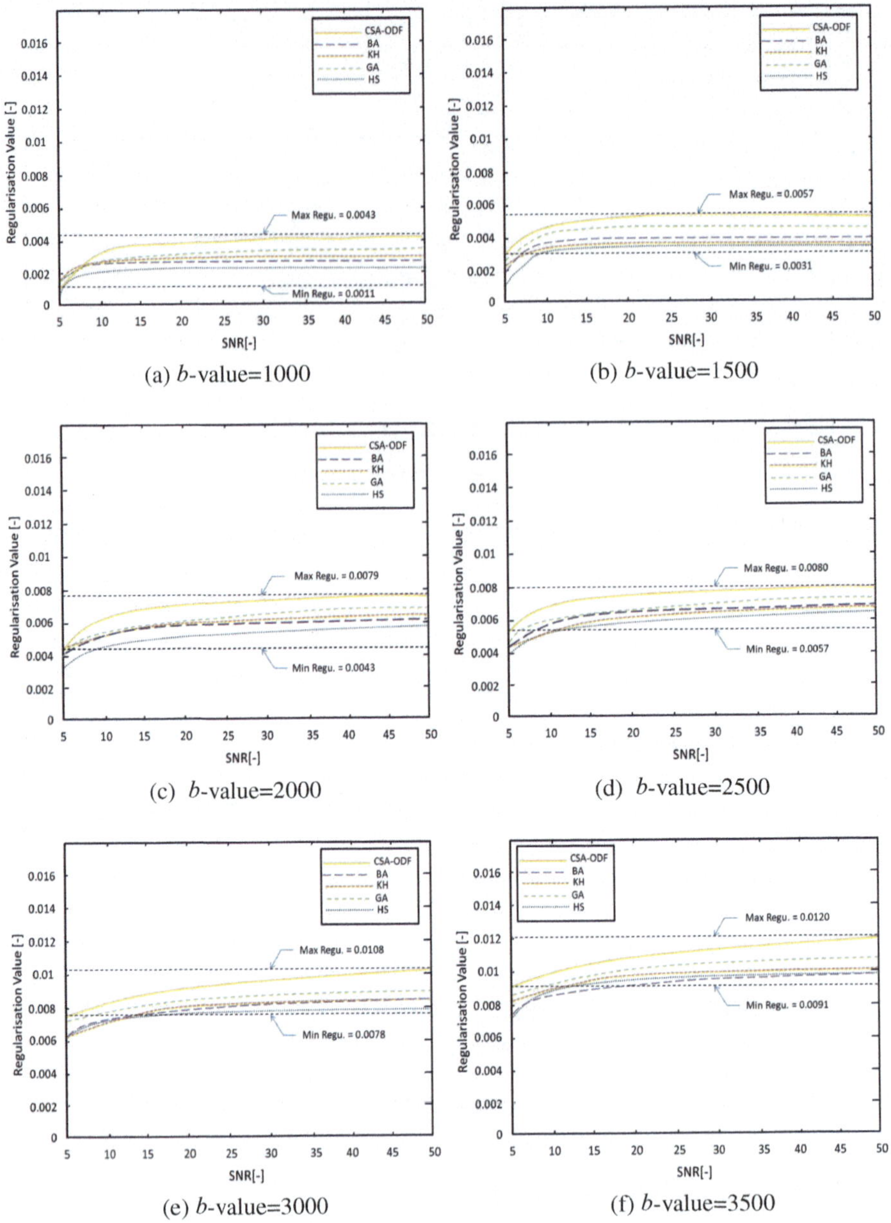

Fig. 5.4 The influence of the SNR and b-value on the HS, KH, BA, GA, and CSA-ODF

This type of evaluation (i.e., influences of SNR and the *b*-value) focuses on demonstrating the robustness of CSA-ODF by exposing it to different SNR values and exhibiting its accuracy using different *b*-value. Moreover, the results of HS, KH, BA, and GA are compared with the result of CSA-ODF to prove the efficiency of the latter. The results aim to show the capability of CSA-ODF and the other algorithms to find the best smoothness (i.e., regularization) of the ODFs. The experimental settings include *b*-value = (1000, 1500, 2000, 2500, 3000, and 3500), SNR = (5, 10, 15, 20, 25, 30, 30, 35, 40, 45, and 50), and an average of 100 run times for each value (Jian et al. 2007; Tuch 2004).

Figure 5.4 shows the results in groups (i.e., subfigures), with each group presenting the results of a fixed SNR and different *b*-value. For example, Fig. 5.4a shows the results of *b*-value = 1000 and SNR = 5, 10, 15, 20, 25, 30, 30, 35, 40, 45, and 50, whereas Fig. 5.4b shows the results of *b*-value = 1500, and so on.

All the results indicate that the regularization value increases by increasing the *b*-value and SNR. Such finding is logical because increasing the *b*-value and SNR will improve the accuracy of the image. That is, improving the smoothness of ODF shape makes the image more accurate. In addition, as shown in Fig. 5.4a–d, the best value of SNR is approximately 30 (i.e., the regularization value achieves the best value when SNR is approximately 30 and then becomes stable afterward). By contrast, Fig. 5.4e, f show that the regularization value continuously increases from the beginning until the end. This finding indicates that the effect of the *b*-value is greater than that of SNR. Finally, all the results show that CSA-ODF clearly outperforms the other algorithms. However, CSA-ODF still requires additional enhancement to achieve better regularization results, and thus, improves the accuracy of images.

5.5 Conclusion

This chapter presents the basic CSA-ODF. The five main steps of CSA-ODF are thoroughly described.

CSA-ODF is evaluated using a synthetic dataset. The angular resolution of CSA-ODF and the influences of the number of initial solutions, *Lévy* flight, SNR, and *b*-value are discussed. Experiments have been performed to examine the efficiency of the CSA-ODF solutions. Moreover, the efficiency of CSA-ODF is proven through comparisons with BA, GA, KH, and HS.

The results demonstrate that although CSA-ODF can obtain viable solutions, these solutions are not impressive because BCSA exhibits weaknesses. This finding may be attributed to the weaknesses in the selection mechanism of CSA-ODF. The weaknesses of CSA-ODF are addressed in the next chapter.

References

Abbasloo, A., Wiens, V., Hermann, M., & Schultz, T. (2016). Visualizing tensor normal distributions at multiple levels of detail. *IEEE Transactions on Visualization and Computer Graphics, 22*(1), 975–984.

Abualigah, L. M., Sawaie, A. M., Khader, A. T., Rashaideh, H., Al-Betar, M. A., & Shehab, M. (2017). β-hill climbing technique for the text document clustering. *New Trends in Information Technology, 60.*

Aganj, I., Lenglet, C., & Sapiro, G. (2010). ODF maxima extraction in spherical harmonic representation via analytical search space reduction. In *International Conference on Medical Image Computing and Computer-Assisted Intervention* (pp. 84–91). Springer.

Alexander, D. C., & Barker, G. J. (2005). Optimal imaging parameters for fiber-orientation estimation in diffusion MRI. *Neuroimage, 27*(2), 357–367.

Bloy, L., & Verma, R. (2008). On computing the underlying fiber directions from the diffusion orientation distribution function. *Medical Image Computing and Computer-Assisted Intervention-MICCAI, 2008*, 1–8.

Descoteaux, M., Deriche, R., Bihan, D. L., Mangin, J.-F, & Poupon, C. (2009). Diffusion propagator imaging: using laplaces equation and multiple shell acquisitions to reconstruct the diffusion propagator. In *Information processing in medical imaging* (pp. 1–13). Springer.

Descoteaux, M., Angelino, E., Fitzgibbons, S., & Deriche, R. (2006). Apparent diffusion coefficients from high angular resolution diffusion imaging: Estimation and applications. *Magnetic Resonance in Medicine, 56*(2), 395–410.

Descoteaux, M., Angelino, E., Fitzgibbons, S., & Deriche, R. (2007). Regularized, fast, and robust analytical q-ball imaging. *Magnetic Resonance in Medicine, 58*(3), 497–510.

Ghosh, A. K. (1977). A numerical analysis of the tensile test for sheet metals. *Metallurgical and Materials Transactions A, 8*(8), 1221–1232.

Jian, B., Vemuri, B. C., Özarslan, E., Carney, P. R., Mareci, T. H. (2007). A novel tensor distribution model for the diffusion-weighted MR signal. *NeuroImage, 37*(1), 164–176.

Laouchedi, M., Megherbi, T., Khabatti, H., Serrat, I., Deriche, Perlbarg, R. V., & Boumghar, F. O. (2014). Odf maxima computation using hill climbing algorithm. In *Proceedings of the 2014 IEEE 11th International Symposium on Biomedical Imaging (ISBI)* (pp. 722–725). IEEE.

Mantegna, R. N. (1994). Fast, accurate algorithm for numerical simulation of levy stable stochastic processes. *Physical Review E, 49*(5), 4677.

Martín, A., Schiavi, E., & de León, S. S. (2016). On 1-laplacian elliptic equations modeling magnetic resonance image rician denoising. *Journal of Mathematical Imaging and Vision*, 1–23.

Scott, A. D., Nielles-Vallespin, S., Ferreira, P. F., McGill, L.-A., Pennell, D. J., & Firmin, D. N. (2016). The effects of noise in cardiac diffusion tensor imaging and the benefits of averaging complex data. *NMR in Biomedicine, 29*(5), 588–599.

Shehab, M., Daoud, M. Sh., AlMimi, H. M., Abualigah, L. M., & Khader, A. T. (2019). Hybridizing cuckoo search algorithm for extracting the ODF maxima in spherical harmonic representation. *International Journal of Bio-Inspired Computation*, (in press).

Shehab, M., Khader, A. T., & Al-Betar, M. A. (2016). New selection schemes for particle swarm optimization. *IEEJ Transactions on Electronics, Information and Systems, 136*(12), 1706–1711. https://doi.org/10.1541/ieejeiss.136.1706.

Shehab, M., Khader, A. T., & Al-Betar, M. A. (2017). A survey on applications and variants of the cuckoo search algorithm. *Applied Soft Computing*.

Shehab, M., Khader, A. T., & Alia, M. A. (2019). Enhancing cuckoo search algorithm by using reinforcement learning for constrained engineering optimization problems. In *2019 IEEE Jordan International Joint Conference on Electrical Engineering and Information Technology (JEEIT)* (pp. 812–816). IEEE.

Shehab, M., Khader, A. T., & Laouchedi, M. (2017). Modified cuckoo search algorithm for solving global optimization problems. In *International Conference of Reliable Information and Communication Technology* (pp. 561–570). Springer.

Shehab, M., Khader, A. T., & Laouchedi, M. (2018). A hybrid method based on cuckoo search algorithm for global optimization problems. *Journal of ICT, 17*(3), 469–491.

Shehab, M., Khader, A. T., &. (2018). Modified cuckoo search algorithm using a new selection scheme for unconstrained optimization problems, *14*, 1.

Shehab, M., Khader, A. T., Al-Betar, M. A., & Abualigah, L. M. (2017). Hybridizing cuckoo search algorithm with hill climbing for numerical optimization problems. In *2017 8th International Conference on Information Technology (ICIT)* (pp. 36–43). IEEE.

Shehab, M., Khader, A. T., Laouchedi, M., & Alomari, O. A. (2018). Hybridizing cuckoo search algorithm with bat algorithm for global numerical optimization. *The Journal of Supercomputing*, 1–28.

Tournier, J.-D., Calamante, F., Gadian, D. G., & Connelly, A. (2004). Direct estimation of the fiber orientation density function from diffusion-weighted MRI data using spherical deconvolution. *NeuroImage, 23*(3), 1176–1185.

Tuch, D. S. (2004). Q-ball imaging. *Magnetic Resonance in Medicine, 52*(6), 1358–1372.

Tuch, D. S., Reese, T. G., Wiegell, M. R., Makris, N., Belliveau, J. W., & Van Wedeen, J. (2002). High angular resolution diffusion imaging reveals intravoxel white matter fiber heterogeneity. *Magnetic Resonance in Medicine, 48*(4), 577–582.

Van Wedeen, J., Hagmann, P., Tseng, W.-Y. I., Reese, T. G., & Weisskoff, R. M. (2005). Mapping complex tissue architecture with diffusion spectrum magnetic resonance imaging. *Magnetic resonance in Medicine, 54*(6), 1377–1386.

Veraart, J., Fieremans, Els., & Novikov, D. S. (2016). Universal power-law scaling of water diffusion in human brain defines what we see with MRI. arXiv:1609.09145.

Chapter 6
Modified Cuckoo Search Algorithm (MCSA) For Extracting the ODF Maxima

6.1 Introduction

In this section, the modified CSA (MCSA) is presented for extracting the ODF maxima (MCSA-ODF). CSA was proposed by Yang and Deb in 2009. To date, work on this algorithm has significantly increased, and CSA currently has its rightful place among other optimization methodologies (Shehab et al. 2017). MCSA is based on replacing the random selection process with the tournament selection scheme. Thus, the probability of achieving better results is increased, thereby avoiding premature convergence. Performance is validated by applying several benchmarks (Shehab and Khader 2018). The results of the experimental indicate that MCSA performs better than the standard CSA and the other compared methods. Subsequently, MCSA is applied to extract the ODF maxima, namely, MCSA-ODF.

BCSA is one of the most effective optimization techniques. It has been successfully tailored for various optimization problems (Stewart et al. 2016; Basu and Chowdhury 2013; Abualigah and Khader 2017). However, it uses random selection to find a solution, which is then compared with the solution of $Lévy$ flight. The use of random selection may lead to the loss of diversity and neglect of the best solutions, which in turn, the premature convergence and quick stagnation without generating efficient results. CSA modification, which involves replacing the random selection scheme with the tournament selection scheme, has been proposed because of the aforementioned reasons (Shehab et al. 2017).

The computational results show that the proposed MCSA exhibit faster convergence and higher computational precision. In addition, it is more effective in searching for the optimal objective function value. The performance of MCSA is evaluated using 13 benchmark functions that have been carefully selected from the literature. The results illustrate the effectiveness of the proposed algorithm compared with the original CSA on most benchmark cases. Then, MCSA is applied to extract the ODF maxima.

The rest of this section is organized as follows: the related works of the MCSA are being described in Sect. 6.2. Section 6.3 presents the tournament selection scheme.

M. Shehab, *Artificial Intelligence in Diffusion MRI*, Studies in Computational Intelligence 877, https://doi.org/10.1007/978-3-030-36083-2_6

The MCSA and MCSA-ODF illustrate in Sect. 6.4. Subsequently, MCSA is evaluated by using benchmark problems and it is applied to extract the maxima of ODF in Sect. 6.5. Finally, the conclusion is shown in Sect. 6.6.

6.2 Related Works of Modified Cuckoo Search Algorithm

MCSA with rough sets was introduced to deal with high-dimensional data through feature selection (MCSAHD) (El Aziz and Hassanien 2016). MCSAHD is utilized to build the fitness function based on two factors. The first factor is the number of features in a reduced set, which indicates that reducing the number of learning parameters leads to a decrease in consumed time and an increase in convergence speed. The second factor is classification quality and increase in diversity (Cao and Jiang 2016). MCSAHD is validated using several public datasets. The experimental results show that MCSAHD is highly efficient in terms of classification performance and more stable than ACO, PSO, BA, and GA.

The short-term hydrothermal scheduling (ST-HTS) problem considers the reservoir volume constraint. Many algorithms, such as SA, have attempted to solve this problem. However, despite resolving the drawbacks, SA has not been widely applied because it requires a long simulation time but achieves a low-quality solution. Accordingly, Dinh et al. (2016) proposed MCSA for the ST-HTS problem (CSA-ST-HTS) by considering the existence of important elements, such as reservoir volume, fuel cost function thermal unit, and power loss in a transmission line Abualigah et al. (2017). In their work, CSA-ST-HTS focused on generating a new CSA solution and maintaining solution generation based on an alien egg discovery. They partitioned all the eggs into high- or low-quality based on the value of their fitness function. The best egg selected will be used to obtain the increased value. The performance of CSA-ST-HTS was validated by testing on two power systems and comparing it with the performance of BCSA and other methods in the literature. The results showed that CSA-ST-HTS achieved better and faster solutions than the other existing methods.

CSA has proven its worth since it was proposed in 2009; it is a robust and efficient tool for yielding global optima (Shehab et al. 2018). The accuracy of CSA primarily depends on the initial solution and its place from the target value. Thus, CSA may cover many generations. However, CSA can lead to a delay in the convergence rate given that evolutionary operators are used in each generation. Therefore, Rao and Venkaiah (2016) proposed MCSAEDM to improve the performance of CSA and optimize the wire-EDM process while machining Inconel-690. The authors utilized a two-stage initialization process to increase the probability of finding the optimal solution. In the first stage, the population is divided into subpopulations, with each subpopulation represented as a vector. The subpopulation vector contains the evaluation of the value of the objective function for each string. Then, the best string is selected based on its fitness function value. Finally, the aforementioned procedures are repeated for all subpopulation vectors. In the second stage, all the best strings are combined to form a new subpopulation vector; then, the evolutionary operators are applied to the newly formed population vector. The robust and accurate performance of MCSAEDM is maintained by testing it on standard benchmark problems Al-Sai and Abualigah (2017). MCSA was also applied to the wire electrical dischargemachining process. It was determined to be faster and more accurate than BCSA, PSO, GA, and response surface methodology.

CSA is introduced to converge to the global optimum (Yang and Deb 2010). However, a fast convergence rate cannot be guaranteed because CSA depends on random walks. Walton et al. (2011b) used MCSARW as a novel robust optimization algorithm by introducing two modifications. The first modification considers *Lévy* flight step size α to increase a wide range of applications and maintain the attractive features of the original method to be closer to the solution. The second modification realizes the addition of exchange information between the eggs in an attempt to slow down convergence to a minimum. In BCSA, no information exchange occurs among individuals. The second modification groups the top eggs, with the second egg in the group to be selected randomly. Thereafter, a new egg is generated based on the aforementioned two eggs. The results exhibited better solutions than BCSA, PSO, and DE using standard optimization benchmark functions. MCSARW has also proven its efficiency on the engineering problem (Walton et al. 2011a).

6.3 Tournament Selection Scheme

Tournament selection is among the most popular selection methods in GAs. It was proposed by Grefenstette and Baker (1989). Algorithm 3 shows the principle of the tournament selection scheme, which starts from the random selection of t individuals from $P_{(t)}$ population and then proceeds to the selection of the best individual from tournament t. This procedure is repeated n times. The best choice is frequently between two individuals, and thus, this scheme is called binary tournament, where the choice is between t individuals, called tournament size (Blickle and Thiele 1995). Figure 6.1 illustrates the steps of the tournament selection scheme. The first step is selecting the solutions randomly and placing them in groups. The capacity of each group is equal to t (i.e., $t = 3$, which indicates three solutions in each group). The best solution or solutions (based on t in the second step) from each group is determined for the second step. The second step is repeated until the best solution is achieved from all the groups. Selection pressure can be easily adjusted by changing tournament size t. A large value of t leads to a weak quality of the solution and low diversity (Melanie 1999; Blickle and Thiele 1995a; Shehab et al. 2019).

Algorithm 3 Pseudo code of the Tournament Selection Scheme

1: **Input:** $The population P(\tau) the tournament size t \in \{1, 2, ..., N\}$
2: **Output:** The population after selection $P(\tau)'$
3: tournament $(t, J_1, ..., J_N)$:
4: **for** $i \leftarrow 1 to N$ **do**
5: $J_i' \leftarrow$ best fit individual out of t randomly picked individuals from $(J_1, ..., J_N)$;
6: **end for**
7: **Return** $(J_1', ..., J_N')$ and select the best;

Fig. 6.1 Representation of tournament selection scheme

Tournament selection has several advantages. When this scheme is implemented in parallel, it exhibits efficient time complexity (Sharma and Wadhwa 2014), does not require fitness scaling or sorting (Goodman 2014), and has low susceptibility to a takeover by dominant individuals (Oladele and Sadiku 2013).

6.4 MCSA and MCSA-ODF

In this section, a novel MCSA that introduces tournament selection into BCSA is presented, and its pros and cons are outlined in its application to benchmark functions. Then, MCSA is used to solve the ODF problem, i.e., MCSA-ODF, which is evaluated by applying it to a synthetic MRI dataset.

In BCSA, the new solution is selected for the first time via *Lévy* flight. Then, another solution is selected randomly. The fitness of the two solutions is compared (Shehab et al. 2018). BCSA has proven its efficiency compared with other meta-heuristic algorithms. However, the use of the random selection method to find solutions has several drawbacks. For example, other potential solutions with high fitness values are neglected; moreover, this method is inefficient in large problems (Shehab et al. 2016). That is, the fittest solutions should have a higher probability to guide the processes of generating new solutions (i.e., Darwin's principle of "survival of the fittest") (Abualigah et al. 2017). Hence, the quality of the solutions will be weak, which leads to slow convergence and instabilities.

MCSA takes advantage of the characteristics of the tournament selection scheme as mentioned earlier to compensate for the drawbacks of BCSA. Algorithm 4 shows how tournament selection has been applied to line 7 of BCSA. MCSA will share the information of the entire existing solutions. In addition, the opportunity for survival of the best solutions with high fitness will increase, thereby avoiding slow convergence but maintaining the principle of "survival of the fittest".

Algorithm 4 :Modified Cuckoo Search Algorithm

1: Objective function $f(X)$;
2: Generate initial population of n host nests;
3: Evaluate fitness and rank eggs;
4: **while** $t < Max\_iterations \ or \ (Stopcrietrion)$ **do**
5:   Get a cuckoo randomly by *Lévy* flights F_i;
6:   evaluate its quality $fitness F_i$;
7:   ***Call the tournament selection (Algorithm 3) to get*** F_j
8:   **if** $F_i > F_j$ **then**
9:     replace current solution with the new solution;
10:   **end if**
11:   New egg (say X_{new})is laid by host bird via *Lévy* flights with a mutation fraction of $P\alpha$;
12:   evaluate its quality $fitness F_{X_{new}}$;
13:   **if** $F_{X_{new}} < F_X$ **then**
14:     replace current solution with the new solution
15:   **end if**
16:   Rank the solutions and find the current best
17: **end while**
18: Postprocess results and visualization

With regard to the ODF problem, the CSA-ODF described in the previous chapter achieves good results. However, these results have weaknesses that should be improved, such as the angular resolution. Reducing angular resolution is recommended to improve the accuracy of an image. In addition, CSA-ODF clearly affects the parameters of BCSA. This phenomenon indicates that BCSA is not sufficiently strong for searches. Therefore, MCSA-ODF is proposed to solve the drawbacks of CSA-ODF by applying MCSA. As shown in Algorithm 5, the pseudo-code of MCSA-ODF illustrates the steps for extracting the ODF maxima.

Notably, modification is realized in line 8, where tournament selection is applied to choose two solutions that are supposed to be the best solutions for Max_i's neighbors. Through this step, MCSA-ODF will improve the quality of the solutions because it is based on the fittest solutions.

Algorithm 5 MCSA-ODF

Objective function $f(X)$
while $t < Max\_iterations$ **do**
  Generate initial population includes SH coordinates $xi_{(i=1,2,...,u)}$, where $x_i = (\theta_i, \phi_i)$
  while stopping criterion **do**
    Get a solution by *Lévy* flights: Max_i /** assume that the best solution**/
    Check if the Max_i in the search space
    Evaluate its quality/ fitness $F(Max_i)$ by Eq.4.5
    Call the tournament selection (Algorithm 3)
    to determine two solutions Max_j ***and*** Max_k /** neighbors of the Max_i **/
    Check if the solutions Max_j and Max_k are in the search space
      Evaluate its quality/ fitness $F(Max_j)$ and $F(Max_k)$ by Eq.4.5
      Get the Minimum and Maximum between Max_i, Max_j, and Max_k : MIN and MAX /** If the MIN pass from the last checking $P\alpha$ the other will pass **/
      Check the MIN by Probability ($P\alpha$) /** if the MIN discover: will be abandoned the solutions and build new solution by *Lévy* flights **/
  end while
  Maximum= Max
  Store the other two solutions (precaution)
  Labeling the points that constitute the vicinity of the detected *Maximum* and its antipodal
end while
Postprocess results and visualization

6.5 Experiments and Results

The main objective of this section is to study the performance of the proposed MCSA. The evaluation of MCSA performance includes two main experiments.

1. **Benchmark optimization problems**:
   Thirteen well-known benchmark functions are used for comparison to test the performance of MCSA. Table 4.2 describes these benchmark functions. The benchmark functions are divided into unimodal and multimodal problems.
2. **Synthetic Data**:
   The synthetic data used to evaluate the performance of MCSA-ODF in detecting the ODF maxima and to check its efficiency by comparing it with other versions of CSA-ODF for use in real data. The settings of the synthetic data are similar to the settings of the data used in the CSA-ODF experiments in Sect. 4.4.3.

6.5.1 Experimental Design

MCSA is compared with five optimization algorithms (i.e., KH, HS, GA, BA, and BCSA) using global optimization problems. In all the experiments, the parameters used for KH, HS, GA, and BA are the same as those used in each original algorithm. BCSA has two other parameters, namely, the number of host nests (population size n) and the probability of discovery $P\alpha$. In Sect. 6.5.2, different settings have been used for each parameter, such as n = (5, 10, 15, 20, 50, 100, 150, 250, and 500) and $P\alpha$ = (0, 0.1, 0.2, 0.3, 0.4, 0.5, 0.6, 0.7, 0.8, 0.9, and 1).

Tables 6.1 and 6.2 show the best normalization (i.e., absolute best minima found by each algorithm) and mean normalization (i.e., average minima found by each algorithm) with different values of parameter n. The best value of n will be fixed and used in the experiments on $P\alpha$. Tables 6.3 and 6.4 show the best normalization and mean normalization with different values of $P\alpha$. Tables 6.5 and 6.6 show the best and mean normalizations for the six algorithms (KH, HS, GA, BA, BCSA, and MCSA). A total of 100 implementations with 10,000 run times are performed for each algorithm on each benchmark function to decrease the influence of randomness. After proving the efficiency of MCSA, MCSA-ODF is applied to the synthetic data, as shown in Sect. 6.5.2(b).

6.5.2 *Experimental Results and Discussion*

6.5.2.1 Benchmark Function

The influence of n is investigated through an array of experiments with n = 5, 10, 15, 20, 50, 100, 150, 250, and 500. As shown in Table 6.1, all the results are close to one another except when $n=20$ (i.e., by looking at the last row (total), the numbers are nearly similar except when $n=20$, the number is 10). The results of the best normalization when n = 20 outperform the others, particularly for F1–F4, F6, F8–F10, F12, and F13. Furthermore, the results of the mean normalization in Table 6.2 refer to the convergence of the results for n = 20 by achieving 11 best results out of 13. Therefore, the value of n will be set to 20 in the experiment on $P\alpha$.

From Table 6.3, MCSA achieves approximately equivalent results for all $P\alpha$ values (the best 3 and 4 of 13 benchmarks are obtained); however, when $P\alpha = 0.2$, MCSA performs the best on 9 out of 13 benchmarks, namely, F1, F3–F7, F10, F12, and F13. By contrast, the worst results are obtained when $P\alpha = 1.0$, when the best on 1 (F3) out of 13 benchmarks is achieved.

In Table 6.4, the result of MCSA is the most effective in finding the objective function minimum when $P\alpha = 0.2$, which obtains 9 out of 13 benchmarks (F1, F3–F5, F7, and F10–F13). In short, MCSA exhibits the best performance in both best and mean normalization when $P\alpha = 0.2$. Hence, $P\alpha$ is set to 0.2 in the present study.

Table 6.5 illustrates the mean normalization of the results. MCSA outperforms the other algorithms by achieving the minimum objective function on 9 out of 13 benchmarks, namely, F1–F4, F6, F8, F9, F12, and F13, and performing the best on 5 out of 13 benchmarks, namely, F1, F2, F4, F9, and F13. MCSA is followed successively by GA, BA, KH, and HS.

Table 6.1 Best normalized optimization results with different n

	n								
	5	10	15	20	50	100	150	250	500
F1	5.41E+01	**1.00**	1.17E+02	**1.00**	1.49E+01	9.91E+02	5.94E+02	1.93E+02	3.79E+02
F2	5.23E+01	7.13E+00	**1.00**	**1.00**	2.65E+00	**1.00**	2.60E+01	5.16E+01	1.70E+01
F3	1.044	**1.00**	1.049	**1.00**	1.171	1.049	1.042	1.046	1.177
F4	**1.00**	**1.00**	**1.00**	**1.00**	**1.00**	**1.00**	**1.00**	**1.00**	**1.00**
F5	2.41E+01	8.37E+00	4.45E+01	1.93E+00	2.19E+01	3.70E+00	1.12E+01	4.90E+01	**1.00**
F6	1.74E+01	6.89E+00	2.39E+01	**1.00**	2.36E+01	6.65E+01	2.16E+01	3.22E+01	1.82E+01
F7	1.23E+00	1.04E+00	**1.00**	1.09E+00	1.09E+00	1.04E+00	1.24E+00	1.14E+00	2.24E+00
F8	1.01E+01	1.20E+01	1.09E+00	**1.00**	2.68E+01	3.80E+00	5.96E+00	1.62E+01	3.41E+01
F9	1.04E+00	2.57E+00	2.27E+00	**1.00**	5.68E+00	1.24E+00	1.10E+00	2.55E+00	2.25E+00
F10	**1.00**	**1.00**	**1.00**	**1.00**	**1.00**	**1.00**	**1.00**	**1.00**	**1.00**
F11	13.35	10.53	1.86	1.86	6.39	28.27	**1.00**	25.81	28.77
F12	3.28E+00	7.58E+00	2.85E+00	**1.00**	2.53E+00	1.05E+00	5.00E+00	**1.00**	8.54E+00
F13	**1.00**	1.32E+00	8.20E+00	**1.00**	**1.00**	**1.00**	**1.00**	1.29E+00	2.56E+00
Total[a]	**3**	**4**	**4**	**10**	**3**	**4**	**4**	**3**	**3**

[a]The total (last row) refers to the summation of all number 1.00, which indicates the minimum normalization value (i.e., when the total is higher, the performance is better)

Table 6.2 Mean normalized optimization results with different n

	n								
	5	10	15	20	50	100	150	250	500
F1	2.83E+01	2.49E+01	1.83E+00	**1.00**	1.66E+01	9.32E+01	1.08E+02	1.09E+02	1.98E+02
F2	3.23E+01	3.87E+01	1.02E+01	**1.00**	1.49E+00	1.28E+00	1.10E+01	1.12E+01	1.60E+01
F3	1.58E+00	1.53E+00	1.23E+00	**1.00**	1.15E+00	1.16E+00	1.20E+00	1.17E+00	1.17E+00
F4	2.23E+00	1.60E+00	1.42E+00	**1.00**	1.11E+00	1.02E+00	**1.00**	1.18E+00	1.24E+00
F5	2.69	2.32	2.54	**1.00**	1.93	1.81	1.87	1.66	1.40
F6	1.02E+02	7.62E+01	**1.00**	**1.00**	7.62E+00	1.52E+01	2.62E+01	2.62E+01	2.71E+01
F7	1.01	1.04	**1.00**	1.06	1.02	1.01	1.02	1.02	1.07
F8	1.14E+00	1.29E+00	1.28E+00	**1.00**	**1.00**	1.29E+00	1.24E+00	1.21E+00	1.38E+00
F9	1.42E+02	1.15E+01	1.16E+01	**1.00**	3.43E+00	2.79E+00	2.01E+00	1.24E+01	1.55E+01
F10	**1.00**	**1.00**	**1.00**	**1.00**	**1.00**	**1.00**	**1.00**	**1.00**	**1.00**
F11	1.32	1.32	**1.00**	1.27	1.35	1.43	1.22	1.21	1.21
F12	1.11E+00	1.14E+00	1.25E+00	**1.00**	1.19E+00	**1.00**	1.10E+00	1.02E+00	1.33E+00
F13	**1.00**	1.02E+00	1.03E+00	**1.00**	**1.00**	1.02E+00	**1.00**	1.04E+00	1.01E+00
Total[a]	**2**	**1**	**4**	**11**	**3**	**2**	**3**	**1**	**1**

[a]The total (last row) refers to the summation of all number 1.00, which indicates the minimum normalization value (i.e., when the total is higher, the performance is better)

Table 6.3 Best normalized optimization results with different $P\alpha$

	$P\alpha$										
	0	0.1	0.2	0.3	0.4	0.5	0.6	0.7	0.8	0.9	1.0
F1	6.89E+00	1.55E+01	**1.00**	1.73E+01	3.01E+01	1.88E+01	3.35E+00	5.18E+00	1.16E+01	2.82E+01	1.87E+01
F2	4.28E+01	**1.00**	8.58E+00	3.38E+01	1.42E+01	1.64E+01	8.59E+00	4.22E+01	6.91E+00	1.23E+01	5.94E+01
F3	**1.00**	**1.00**	**1.00**	**1.00**	**1.00**	**1.00**	**1.00**	**1.00**	**1.00**	**1.00**	**1.00**
F4	1.15E+00	1.02E+00	**1.00**	1.08E+00	1.17E+00	1.06E+00	**1.00**	**1.00**	1.23E+00	1.14E+00	1.22E+00
F5	1.06E+00	4.23E+00	**1.00**	3.31E+01	5.11E+00	7.01E+00	5.17E+00	5.39E+00	1.82E+01	**1.00**	1.81E+01
F6	6.36E+00	2.54E+01	**1.00**	6.99E+00	1.13E+01	2.55E+00	7.63E+00	5.63E+00	1.40E+00	1.94E+00	7.95E+00
F7	106.24	106.69	**1.00**	106.32	106.62	106.43	106.57	106.89	106.33	105.31	1.17E+00
F8	5.84E+01	3.46E+00	1.45E+01	2.18E+00	5.15E+00	1.10E+01	6.24E+00	3.67E+00	1.26E+01	**1.00**	1.90E+01
F9	5.06E+01	4.85E+00	1.43E+00	1.95E+01	**1.00**	9.52E+00	6.84E+01	3.28E+01	1.56E+01	2.08E+01	2.55E+01
F10	**1.00**	**1.00**	**1.00**	**1.00**	**1.00**	1.21E+00	**1.00**	**1.00**	**1.00**	1.01E+00	2.12E+00
F11	28.81	**1.00**	2.22	39.71	14.65	29.39	10.08	4.47	12.14	68.27	2.93
F12	4.61E+01	6.03E+00	**1.00**	2.02E+01	6.82E+00	2.54E+00	1.13E+01	4.86E+00	7.52E+00	1.12E+00	1.48E+01
F13	**1.00**	**1.00**	**1.00**	**1.00**	**1.00**	**1.00**	4.50E+00	4.50E+00	4.50E+00	4.50E+00	9.08E+00
Total[a]	**3**	**5**	**9**	**3**	**4**	**2**	**3**	**3**	**2**	**3**	**1**

[a]The total (last row) refers to the summation of all number 1.00, which indicates the minimum normalization value (i.e., when the total is higher, the performance is better)

Table 6.4 Mean normalized optimization results with different $P\alpha$

	$P\alpha$										
	0	0.1	0.2	0.3	0.4	0.5	0.6	0.7	0.8	0.9	1.0
F1	1.20	1.26	**1.00**	**1.00**	1.14	1.02	1.35	1.38	1.32	1.26	1.55
F2	1.29E+00	**1.00**	1.32E+00	4.54E+01	5.25E+01	1.57E+01	8.78E+01	4.42E+01	9.21E+01	1.01E+01	5.18E+01
F3	**1.00**	1.10E+00	**1.00**	1.07E+00	1.05E+00	1.14E+00	1.13E+00	1.25E+00	1.24E+00	1.16E+00	**1.00**
F4	1.11E+00	**1.00**	**1.00**	1.32E+00	1.24E+00	1.20E+00	1.52E+00	1.38E+00	1.52E+00	1.78E+00	1.16E+00
F5	**1.00**	1.06	**1.00**	1.24	1.03	1.27	1.10	1.09	1.13	1.34	1.25
F6	1.10E+01	1.38E+01	1.38E+00	1.38E+00	**1.00**	1.22E+00	1.38E+01	5.53E+01	5.53E+01	2.76E+01	3.88E+01
F7	2.33	2.31	**1.00**	2.37	2.39	2.32	2.33	2.32	2.34	2.31	2.38
F8	4.44E+00	1.25E+00	1.04E+00	**1.00**	1.05E+00	1.19E+00	1.21E+00	1.28E+00	1.29E+00	1.25E+00	1.40E+01
F9	8.13E+00	**1.00**	1.48E+00	9.72E+01	2.40E+01	4.28E+01	3.77E+01	4.69E+01	4.93E+01	4.98E+01	5.48E+01
F10	1.02E+00	**1.00**	**1.00**	**1.00**	**1.00**	**1.00**	**1.00**	**1.00**	**1.00**	1.02E+00	6.65E+00
F11	1.22	1.16	**1.00**	1.18	1.23	1.01	1.02	1.13	1.25	1.07	1.04
F12	1.47E+01	1.22E+00	**1.00**	1.12E+00	**1.00**	1.14E+00	1.25E+00	1.37E+00	1.25E+00	1.19E+00	6.42E+01
F13	**1.00**	1.01E+00	**1.00**	**1.00**	1.04E+00	1.01E+00	4.49E+01	4.49E+01	4.49E+01	4.49E+01	4.35E+01
Total[a]	**3**	**4**	**9**	**4**	**3**	**1**	**1**	**1**	**1**	**0**	**1**

[a]The total (last row) refers to the summation of all number 1.00, which indicates the minimum normalization value (i.e., when the total is higher, the performance is better)

Table 6.5 Mean normalized optimization results

	BA	GA	HS	KH	BCSA	MCSA
F1	4.01E+02	9.11E+01	4.86E+02	**1.00**	**1.00**	**1.00**
F2	1.74E+02	1.08E+01	4.89E+01	8.97E+02	**1.00**	**1.00**
F3	3.88E+00	4.81E+00	1.16E+02	7.56E+00	7.19E+01	**1.00**
F4	8.77E+02	6.54E+00	5.92E+01	3.74E+01	**1.00**	**1.00**
F5	9.76E+00	**1.00**	7.21E+00	8.39E+01	3.09E+00	1.13E+00
F6	6.28E+01	1.99E+01	1.18E+02	5.19E+01	1.05E+01	**1.00**
F7	2.09 E+02	**1.00**	5.05 E+01	2.39 E+01	1.65 E+02	2.25E+00
F8	9.43E+01	8.48E+00	2.38E+01	3.09E+01	3.66E+00	**1.00**
F9	2.66E+02	1.17E+01	5.60E+02	9.14E+01	**1.00**	**1.00**
F10	6.03E+01	**1.00**	1.73E+02	4.01E+00	1.25E+01	3.03E+00
F11	**1.00**	5.9564	8.5483	6.4578	3.7038	1.2457
F12	3.45E+02	5.73E+01	4.67E+02	8.93E+01	1.01E+00	**1.00**
F13	**1.00**	**1.00**	**1.00**	**1.00**	**1.00**	**1.00**
Total[a]	**2**	**4**	**1**	**2**	**5**	**9**

[a]The total (last row) refers to the summation of all number 1.00, which indicates the minimum normalization value (i.e., when the total is higher, the performance is better)

Table 6.6 shows that MCSA performs the best on 11 out of 13 benchmarks, which are F1–F4, F6, and F8–F13. BCSA ranks second and performs the best on F1, F2, F4, F8, F10, and F13. MCSA is followed by GA, which performs best on F5, F7, F10, and F13. KH comes next, with the best performance on F1 and F13. BA and HS rank last.

From the results presented in Tables 6.3 and 6.4, MCSA does not achieve the best results in certain benchmark functions, such as F5 and F7, because these benchmark functions are unimodal functions that are focused on the local optima. Nevertheless, MCSA achieves the best results in other unimodal functions, such as F9, F11, F12, and F13, because the equations in these functions are not as complicated as those in F5 and F7.

Figure 6.2a–h illustrate the most representative convergent curves. The values in the figures are the mean function optimum, which are the true values.

Figure 6.2a–d show that MCSA is capable of finding better solutions compared with all the other methods. In Fig. 6.2a, d, MCSA achieves the best solutions from the 15th and 7th generations, respectively, until the end. In both figures, BCSA ranks second and is followed by GA. By contrast, in Fig. 6.2b, c, MCSA achieves the best results from beginning to end, but exhibits a weakness in the middle generations, such as from the 12th to the 34th generations. In Fig. 2.6b, GA and BCSA perform better than MCSA. Moreover, in Fig. 2.6c, the results of MCSA are nearly similar to the results of BCSA from the 15th to the 26th generations.

Figure 6.2e–h show that MCSA and GA exhibit close performances, with a preference for MCSA, particularly at the end of generations. The ranking in Fig. 6.2e illustrates that the best algorithm is GA, followed by MCSA, and then BCSA. In Fig. 6.2f,

Table 6.6 Best normalized optimization results

	BA	GA	HS	KH	BCSA	MCSA
F1	3.9532	1.5272	4.8696	**1.00**	**1.00**	**1.00**
F2	16.0887	4.3422	27.7036	7.4725	**1.00**	**1.00**
F3	2.41E+2	2.2193	80.9202	5.8517	3.6356	**1.00**
F4	7.11E+01	2.11E+00	1.90E+01	3.75E+01	**1.00**	**1.00**
F5	1.84E+02	**1.00**	2.45E+02	2.2E+01	3.2187E+00	2.53E+00
F6	1.09E+01	34.6501	4.28E+01	3.8978	8.2575E+00	**1.00**
F7	11.0863	**1.00**	7.4952	6.2274	1.9896	2.95E+00
F8	107.4910	3.0915	56.0259	13.3735	**1.00**	**1.00**
F9	22.169	12.5346	138.4713	15.3128	17.0757	**1.00**
F10	13.2849	**1.00**	4.8184	2.3130	**1.00**	**1.00**
F11	12.8324	6.1904	24.5390	10.0382	4.2662	**1.00**
F12	86.1546	8.4855	15.8124	7.7168	1.2473	**1.00**
F13	11.8313	**1.00**	8.5479	**1.00**	**1.00**	**1.00**
Total[a]	**0**	**3**	**0**	**2**	**6**	**11**

[a]The total (last row) refers to the summation of all number 1.00, which indicates the minimum normalization value (i.e., when the total is higher, the performance is better)

MCSA exhibits the best performance, followed by GA. Notably, GA achieves better results than MCSA from the beginning until the 47th generation. Moreover, Fig. 6.2g shows that the results of MCSA and GA are nearly similar in the last five generations with MCSA exhibiting a slight superiority. However, BCSA outperforms MCSA and GA from the beginning until the 18th generation, followed by MCSA, which is outperformed in the 19th and 20th generations. Meanwhile, the priority of GA ranges from generations 21th–44th. Finally, Fig. 6.2h shows that MCSA achieves the best results with an evident difference from the other algorithms, particularly from the beginning until generation 22. Thereafter, the results of MCSA became became extremely close to those of BCSA and GA, with MCSA slightly outperforming the other two.

The results in the preceding two paragraphs are expected, where the first paragraph presents the results of multimodal functions, i.e., Fig. 6.2a–d. As mentioned earlier, multimodal functions frequently contain multiple global and local optima; therefore, the performance of MCSA is good with a few weaknesses that should be improved. The second paragraph presents the results of unimodal functions, i.e., Fig. 6.2e–h. As mentioned earlier, unimodal functions contain only one local optimum; nevertheless, the performance of MCSA is occasionally slightly better than those of the other algorithms.

An analysis of Fig. 6.2a–h indicates that MCSA outperforms the other algorithms, with the occurrence of several weaknesses, particularly in the local optima.

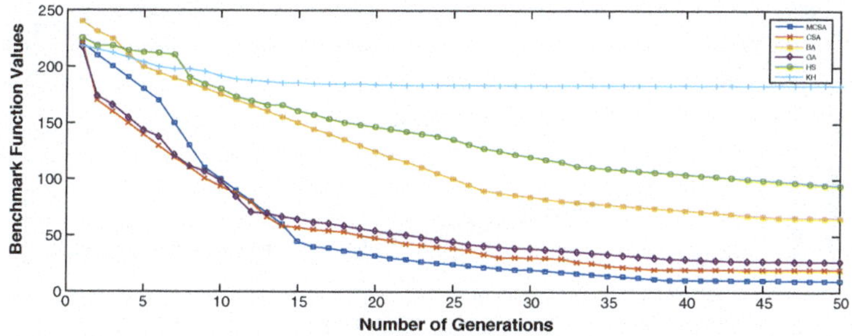

(a) Ackley Function

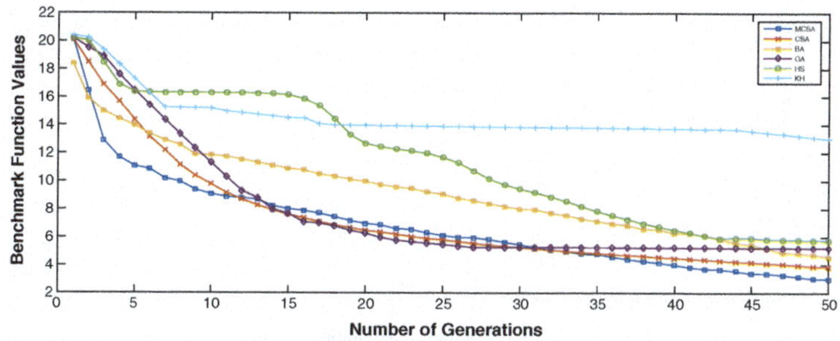

(b) Penalty #1 Function

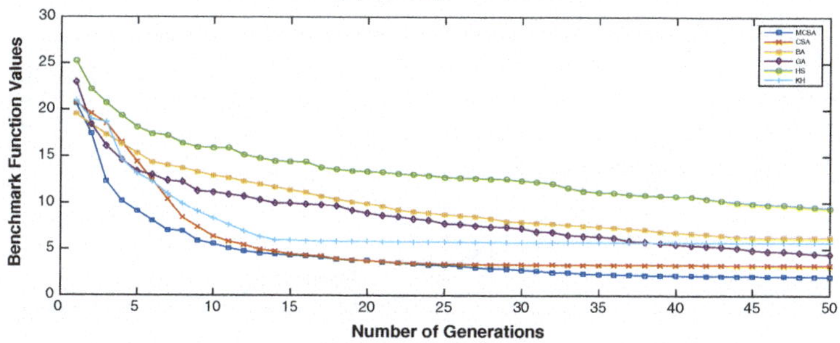

(c) Penalty #2 Function

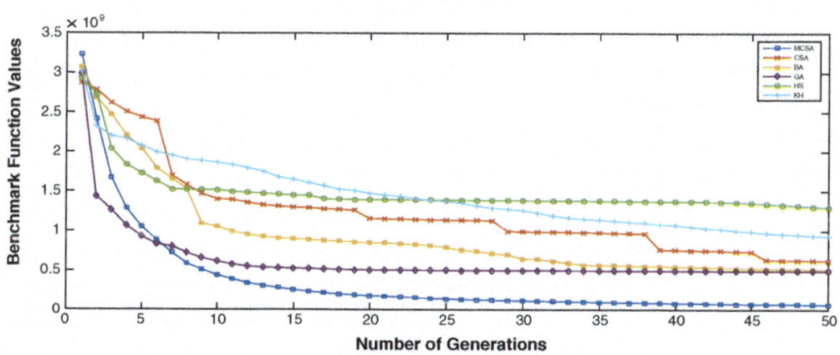

(d) Rastrigin Function

Fig. 6.2 Performance comparison for the benchmark functions

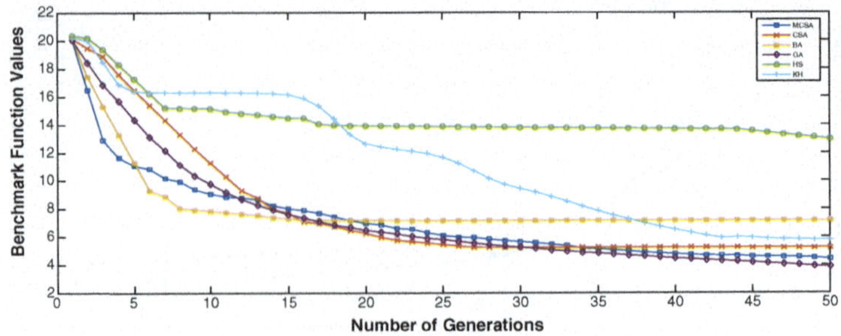

(e) Rosenbrock Function

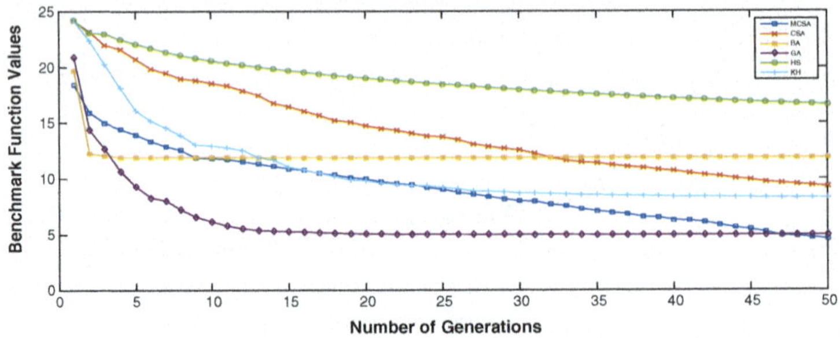

(f) Schwefel 1.2 Function

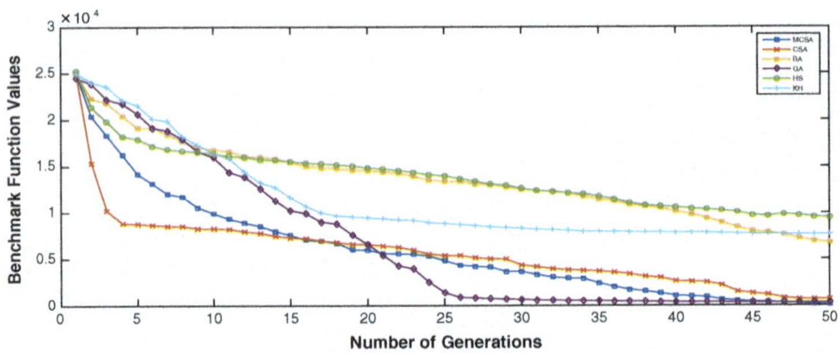

(g) Schwefel 2.22 Function

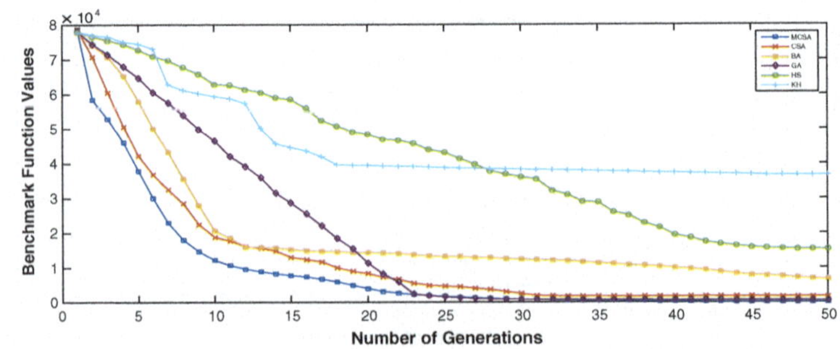

(h) Sphere Function

Fig. 6.2 (continued)

6.5.2.2 Synthetic Data

- **Angular resolution of the MCSA-ODF**
  This experiment is performed to determine the angular resolution of MCSA-ODF. The results are compared with the previous version (i.e., CSA-ODF). Descoteaux (2010)'s conditions are reproduced to reconstruct the ODF in the fourth RSSH as mentioned previously in the description of the synthetic data in Sect. 4.3.1.

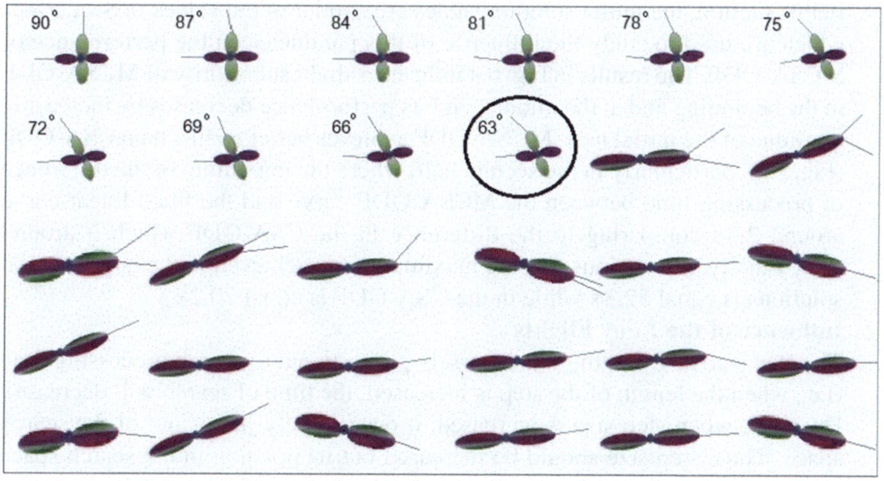

Fig. 6.3 The ability of the MCSA-ODF to detect maxima separated by 63°

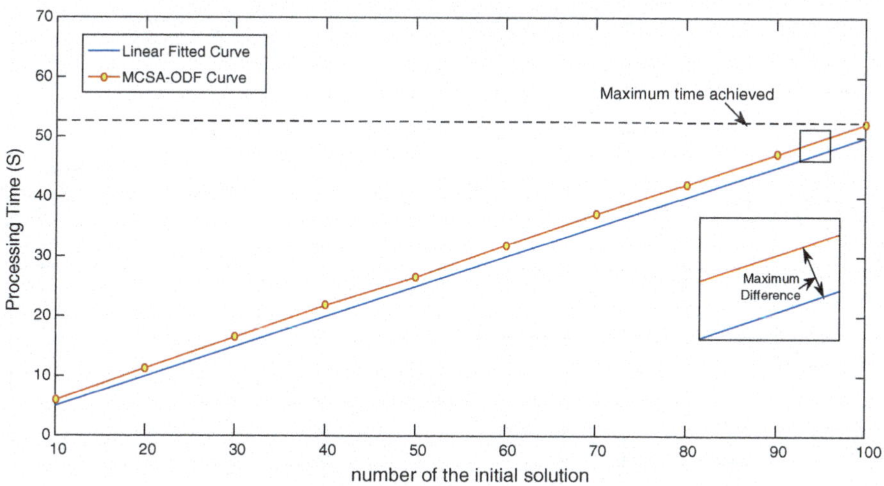

Fig. 6.4 Influence of the number of initial solutions on the processing time of the MCSA-ODF

Figure 6.3 illustrates the ODFs. The ODF in the upper left corner presents a separation angle of 90° between its maxima. This angle is decreased by 3° each time to produce the remaining ODFs. The figure shows that two maxima are detected for separation angles equal to and higher than 63°. This limit corresponds to the circled ODF in the figure. MCSA-ODF achieves better separation angle than CSA-ODF (66°). However, further enhancements should be implemented to obtain better results with small angles.

- **Influence of control parameter**

1. **Influence of the number of the initial solutions**
   In this section, the initial solution values are similar to the values in Sect. 5.4.2, which are used to study the influence of this parameter on the performance of MCSA-ODF. The results in Fig. 6.4 indicate a slight superiority of MCSA-ODF in the beginning and at the middle, and its performance decreases by increasing the value of the initial nest. MCSA-ODF achieves better results than CSA-ODF (Fig. 5.2), particularly in the second half, where the maximum of the difference of processing time between the MCSA-ODF curve and the fitted linear curve around 2.3 s comparing to the difference in the CSA-ODF which it around 4.2 s. Finally, it's obvious that the maximum time achieved in the largest initial solutions is equal 52.3 s while in the CSA-ODF is equal 70.2 s.
2. **Influence of the *Lévy* Flights**
   The step size of *Lévy* flight is inversely proportional to search processing time (i.e., when the length of the step is increased, the time of search will decrease). However, when step size is increased, it occasionally jumps out of the search space. Thus, step size should be increased but its position in the search space should be maintained. Figure 6.5 presents the results of the effect of step size on

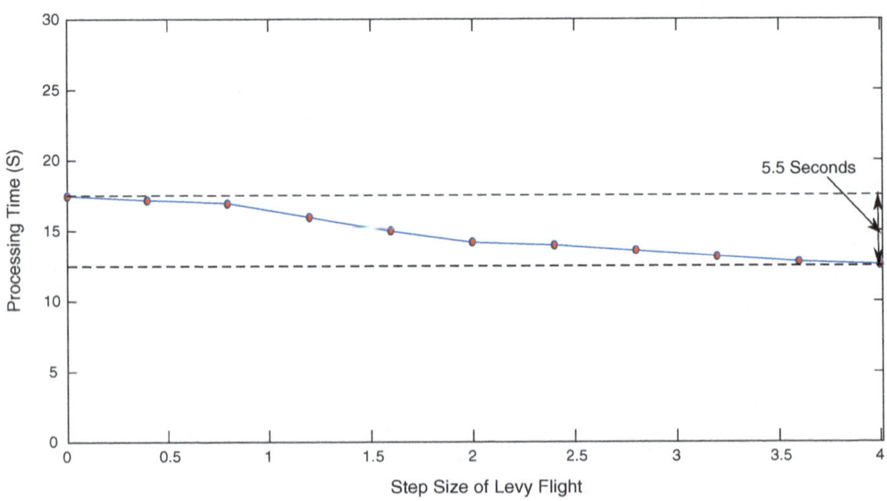

Fig. 6.5 Influence of the step size of *Lévy* Flights on the processing time of the MCSA-ODF

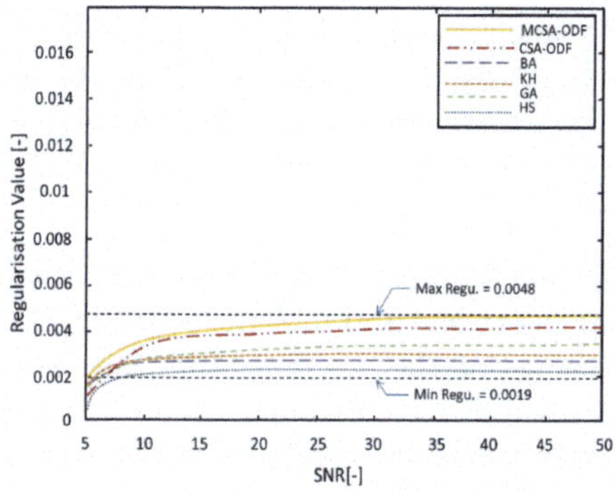

(a) b=1000 s/mm^2

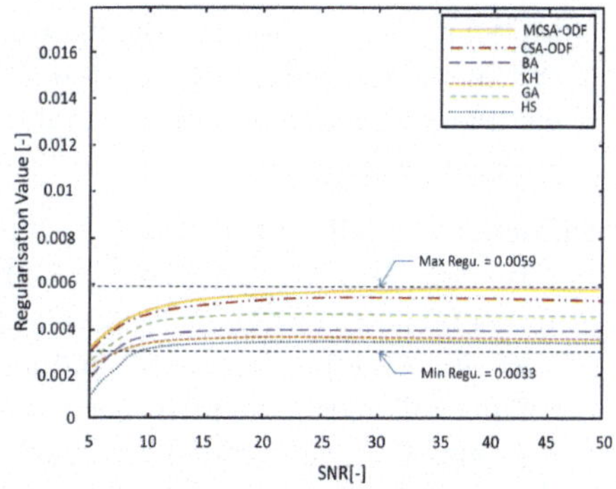

(b) b=1500 s/mm^2

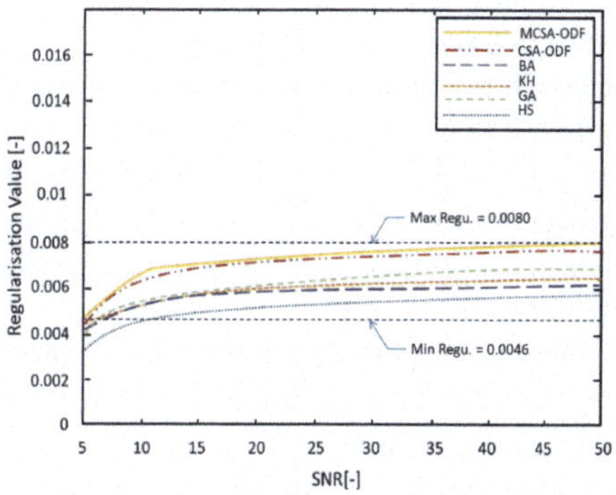

(c) b=2000 s/mm^2

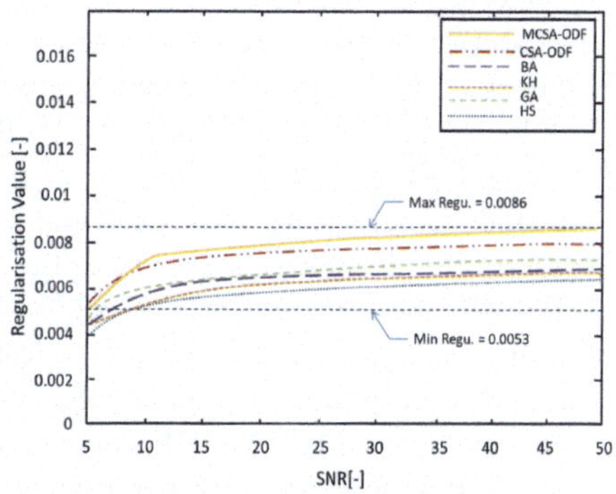

(d) b=2500 s/mm^2

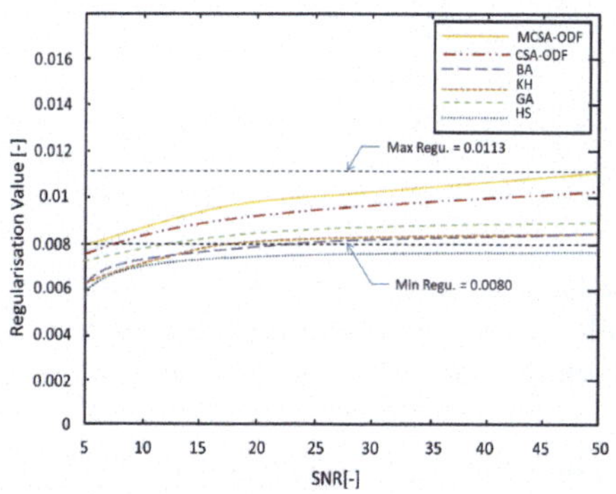

(e) b=3000 s/mm^2

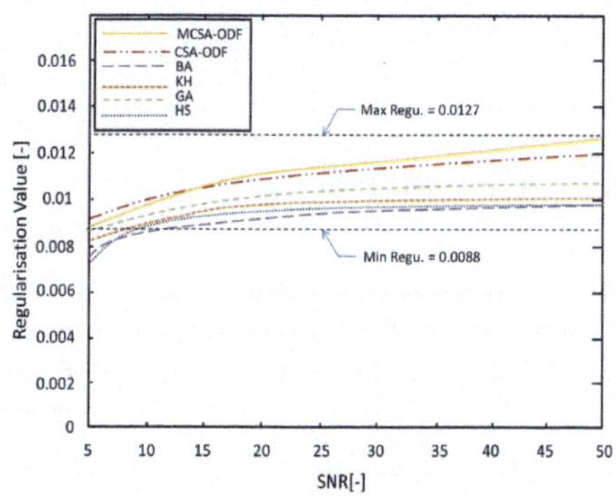

(f) b=3500 s/mm^2

Fig. 6.6 The influence of the SNR and b-value on the HS, KH, BA, GA, CSA-ODF, and MCSA-ODF

MCSA-ODF. As shown in the figure, the time difference between the minimum and maximum values of the step size is nearly 5.5 s, whereas in CSA-ODF, the time difference is approximately 7.5 s, which indicates that MCSA-ODF is affected by step size.

- **Influence of SNR and b-value**:
  In this study, HS, KH, BA, GA, CSA-ODF, and MCSA-ODF are compared using different *b*-values (1000, 1500, 2000, 2500, 3000, and 3500), SNR (5, 10, 15, 20, 25, 30, 30, 35, 40, 45, and 50), and an average of 100 run times for each value (Shehab et al. 2019). As mentioned in Sect. 5.4.2(c), the experiments are divided into groups (i.e., subfigures), and each group shows the results of the effects of SNR and the *b*-value to evaluate the robustness and performance of each algorithm by determining the value of the regularization parameter, which is responsible for the exterior shape of the ODF. Each *b*-values is fixed for all SNR values.
  Figure 6.6 shows groups of subfigures (Fig. 6.6a–f), and each group has a fixed *b*-value and different SNR values. All the results are logical, such that when you increase SNR and the *b*-value, image quality is enhanced (i.e., the smoothness of the ODF is increased). This result can be observed through the increasing regularization value from Fig. 6.6a to f i.e., the maximum regularization value in Fig. 6.6a is 0.0048, and it increases to 0.0127 in Fig. 6.6f.
  Finally, MCSA-ODF has proven the robustness of its performance in these experiments by outperforming other algorithms to find the maximum regularization value. However, MCSA-ODF also has weaknesses, as shown in Fig. 6.6d, f, where CSA-ODF is superior in the beginning (i.e., when SNR = 10 and 15 or less). Therefore, MCSA-ODF should be improved to achieve better regularization values that will increase image accuracy.

6.6 Conclusion

In this chapter, a new variation of BCSA based on a new selection scheme is proposed. MCSA applies the natural selection principle of survival of the fittest to generate a new BCSA. This variation is focused on finding better solutions from the solution space.

The first part of the experiments was conducted using a set benchmark functions that is widely used in the literature. The experimental results of MCSA show that incorporating the new selection scheme (i.e., the tournament selection scheme) into the solution space avoids slow convergence with a long time requirement without effective results due to the mechanism of the tournament selection scheme, which provides a chance for the solution with high fitness to be selected.

The second part of the experiments was conducted using synthetic data. Three approaches have been used to evaluate MCSA-ODF (i.e., angular resolution, influences of the control parameters, and influences of SNR and the *b*-value). The results show that the performance is fairly acceptable compared with those of HS, KH, BA, GA, and CSA-ODF.

Although MCSA achieves good results in most of the benchmark functions, it cannot achieve the best results in all the benchmark functions, particularly those belonging to the local optima (i.e., unimodal functions). Moreover, MCSA-ODF still has weaknesses that should be improved, such as determining a smaller angular resolution, increasing its robustness in dealing with the effects of BCSA parameters, and achieving better regularization value.

References

Abualigah, L. M., Khader, A. T., Hanandeh, E. S., & Gandomi, A. H. (2017). A novel hybridization strategy for krill herd algorithm applied to clustering techniques. *Applied Soft Computing, 60*, 423–435.

Abualigah, L. M., Sawaie, A. M., Khader, A. T., Rashaideh, H., Al-Betar, M. A., & Shehab, M. (2017) β-hill climbing technique for the text document clustering. *New Trends in Information Technology, 60*.

Abualigah, L. M., & Khader, A. T. (2017). Unsupervised text feature selection technique based on hybrid particle swarm optimization algorithm with genetic operators for the text clustering. *The Journal of Supercomputing, 73*(11), 4773–4795.

Al-Sai, Z. A., & Abualigah, L. M. (2017). Big data and e-government: a review. In *2017 8th International Conference on Information Technology (ICIT)* (pp. 580–587). IEEE.

Basu, M., & Chowdhury, A. (2013). Cuckoo search algorithm for economic dispatch. *Energy, 60*, 99–108.

Blickle, T., & Thiele, L. (1995). A comparison of selection schemes used in genetic algorithms.

Blickle, T., & Thiele, L. (1995). A mathematical analysis of tournament selection. In *ICGA* (pp. 9–16). Citeseer.

Cao, B., Jiang, Z. (2016). Increasing diversity and controlling bloat in linear genetic programming. In *2016 3rd International Conference on Information Science and Control Engineering (ICISCE)* (pp. 414–419). IEEE.

Descoteaux, M. (2010). *High angular resolution diffusion MRI: from local estimation to segmentation and tractography*. Ph.D. thesis, INRIA Sophia Antipolis, France.

Dinh, B. H., Nguyen, T. T., & Vo, D. N. (2016). Adaptive cuckoo search algorithm for short-term fixed-head hydrothermal scheduling problem with reservoir volume constraints. *International Journal of Grid and Distributed Computing, 9*(5), 191–204.

El Aziz, M. A., & Hassanien, A. E. (2016). Modified cuckoo search algorithm with rough sets for feature selection. *Neural Computing and Applications*, 1–10.

Goodman, E. D. (2014). Introduction to genetic algorithms. In *Proceedings of the Companion Publication of the 2014 Annual Conference on Genetic and Evolutionary Computation* (pp. 205–226). ACM.

Grefenstette, J. J., & Baker, J. E. (1989). How genetic algorithms work: A critical look at implicit parallelism. In *Proceedings of the third international conference on Genetic algorithms* (pp 20–27). Morgan Kaufmann Publishers Inc.

Melanie, M. (1999). An introduction to genetic algorithms. *Cambridge, Massachusetts London, England, Fifth printing, 3*, 62–75.

Oladele, R. O., & Sadiku, J. S. (2013). Genetic algorithm performance with different selection methods in solving multi-objective network design problem. *International Journal of Computer Applications, 70*(12)

Rao, M. S., & Venkaiah, N. (2016). A modified cuckoo search algorithm to optimize wire-EDM process while machining inconel-690. *Journal of the Brazilian Society of Mechanical Sciences and Engineering*, 1–15.

Sharma, P., & Wadhwa, A. (2014). Analysis of selection schemes for solving an optimization problem in genetic algorithm. *International Journal of Computer Applications*, *93*(11).

Shehab, M., & Khader, A. T. (2018). Modified cuckoo search algorithm using a new selection scheme for unconstrained optimization problems. *14*, 1.

Shehab, M., Daoud, M. Sh., AlMimi, H. M., Abualigah, L. M., & Khader, A. T. (2019). Hybridizing cuckoo search algorithm for extracting the ODF maxima in spherical harmonic representation. *International Journal of Bio-Inspired Computation*, (in press).

Shehab, M., Khader, A. T., & Al-Betar, M. A. (2016). New selection schemes for particle swarm optimization. *IEEJ Transactions on Electronics, Information and Systems, 136*(12), 1706–1711. https://doi.org/10.1541/ieejeiss.136.1706.

Shehab, M., Khader, A. T., & Al-Betar, M. A. (2017). A survey on applications and variants of the cuckoo search algorithm. *Applied Soft Computing*.

Shehab, M., Khader, A. T., & Alia, M. A. (2019) Enhancing cuckoo search algorithm by using reinforcement learning for constrained engineering optimization problems. In *2019 IEEE Jordan International Joint Conference on Electrical Engineering and Information Technology (JEEIT)* (pp. 812–816). IEEE.

Shehab, M., Khader, A. T., & Laouchedi, M. (2017). Modified cuckoo search algorithm for solving global optimization problems. In *International Conference of Reliable Information and Communication Technology* (pp. 561–570). Springer.

Shehab, M., Khader, A. T., & Laouchedi, M. (2018). A hybrid method based on cuckoo search algorithm for global optimization problems. *Journal of ICT, 17*(3), 469–491.

Shehab, M., Khader, A. T., Laouchedi, M., & Alomari, O. A. (2018). Hybridizing cuckoo search algorithm with bat algorithm for global numerical optimization. *The Journal of Supercomputing*, 1–28.

Stewart, B., Wild, C. P. et al. (2016). World cancer report 2014. *World*.

Walton, S., Hassan, O., & Morgan, K. (2011). Using proper orthogonal decomposition to reduce the order of optimization problems. In *Proceedings of the 16th International Conference on Finite Elements in Flow Problems, Munich* (p. 90).

Walton, S., Hassan, O., Morgan, K., & Brown, M. R. (2011b). Modified cuckoo search: A new gradient free optimisation algorithm. *Chaos, Solitons & Fractals*, *44*(9), 710–718.

Yang, X.-S., & Deb, S. (2010). Engineering optimisation by cuckoo search. *International Journal of Mathematical Modelling and Numerical Optimisation*, *1*(4), 330–343.

Chapter 7
Hybridization Cuckoo Search Algorithm for Extracting the ODF Maxima

7.1 Introduction

BCSA has constantly attracted the interest of investigators from diverse disciplines worldwide since its introduction in 2009. This interest has led to various hybridizations for improving the performance of the basic BCSA. These hybridizations can improve MCSA and achieve favorable results.

The remainder of this chapter is organized as follows. Section 7.2 discusses the related works on the hybridization of the cuckoo search algorithm. Two types of hybridization are presented. Section 7.3 describes the population-based algorithm that contains MCSA with BA (CSBA). Section 7.4 presents the local search-based algorithm that includes CSBA with HC (CSAHC). Section 7.5 discusses the comparison summary of all versions. Section 7.6 draws the conclusion.

7.2 Related Work of Hybridization Cuckoo Search Algorithm

The optimal design of water distribution systems was solved by using a hybrid optimization method, which combined the BCSA and a HS (CSAHS) (Sheikholeslami et al. 2016). In the CSAHS, the explorative stages of searching and finding the promising regions of the search space were conducted by BCSA. Then, the hybridization between BCSA and HS algorithms was utilized in the exploitation phase. CSAHS was improved through harmony memory (HM), which consists of information extracted online through a search and the addition of the pitch-adjusting operation of HS through the evolution process. Thus, HM vectors become the BCSA population, and an evolving process is performed in the BCSA. In addition, CSAHS was improved by utilizing a self-adaptive strategy for the dynamic adaption of the control parameters

M. Shehab, *Artificial Intelligence in Diffusion MRI*, Studies in Computational Intelligence 877, https://doi.org/10.1007/978-3-030-36083-2_7

using a learning mechanism. The performance of the CSHS was demonstrated by employing four water distribution system design problems with increased scales and complexity. The results illustrated that the CSHS method overtook the basic BCSA, proved its efficiency to obtain optimal solutions, and enhanced the convergence rate (Shehab et al. 2017a).

Wang et al. (2016b) proposed a hybrid algorithm that combined BCSA and a HS (HS/CSA) for continuous optimization problems. In the HS/CSA method, the pitch adjustment of HS was used to update the process of the BCSA, which leads to the increase of population diversity. The improved elitism scheme was used to retain the best individuals in the cuckoo population as well. The performance of HS/CSA was evaluated by means of testing the set of benchmark functions. The obtained results showed that the HS/CSA achieved better outcomes in comparison with ACO, PSO, GA, HS, DE, and basic BCSA.

An early diagnosis of a disease is an important issue in life science and is effective in avoiding the risk to human health. Liu and Fu (2014) proposed a disease diagnosis model based on cuckoo search, PSO, and support vector machine (CSA-POS-SVM). The authors used SVM to solve classification and regression problem on a dataset that is nonlinearly separable and handle data that are not fully linearly separable Reyaz-Ahmed et al. (2009). Additionally, the PSO was used to make use of particles moving in the n-dimensional space to search for solutions to the functions of n-variables. This allows the particles to distribute in the problem space by attracting the particles of the best solutions. Thus, it is initialized with a group of random particles (solutions) and then searches for optima by updating each generation (Ardjani et al. 2010). BCSA discovered that the random-walk style search has better performance for *Lévy* flights rather than the simple random walk. BCSA with its random-walk style is also simple in comparison with other algorithms. Experimental results showed that the CS-PSO-SVM realized better classification accuracy and F-measure than GASVM and PSO-SVM; thus, it is efficient in disease diagnosis.

The linear least squares problem solved by hybridization algorithm between Newton method (NM) and BCSA is called CSANM (Abdel-Baset and Hezam 2016b). The authors benefited from BCSA for fast convergence and global search as well as from NM for the ability of strong local search. The experimental results showed the convergence efficiency and computational accuracy of the CSANM in comparison with the basic BCSA and HS based on NM (HSNM).

Quadratic assignment problems (QAPs) are considered to be NP-hard problems, which cannot be easily solved by exact methods. Therefore, Dejam et al. (2012) proposed a hybrid algorithm combined with the BCSA of TS (i.e., CSA-TS) to solve QAPs. In their research, the QAPs were initially tackled using BCSA. Thereafter, these were combined with TS, which focused on the local search to increase the optimization precision. The experimental results indicated that the proposed algorithm performs better than ABC and GA.

7.3 Hybridization of MCSA with Bat Algorithm

MCSA is a favorable optimization technique and has been successfully designed to solve global optimization problems (Shehab et al. 2017c). However, this technique has drawbacks, such as utilization of global search; in certain cases, this technique may surround local optima. $L\acute{e}vy$ flight is responsible for determining the search, although results cannot be guaranteed if the step size is significantly large. Thus, the algorithm has a slow convergence rate (Walton et al. 2011b). Searches are also performed independently in MCSA. In particular, no information exchange occurs among individuals Long (2014).

MCSA uses $P\alpha$ for the last check of the best solution (i.e., by nature, a measured match ratio between the eggs of the host and the cuckoo) after selecting the best solution. The best solution should be selected carefully, and being trapped in a local optimum should be avoided to ensure that the step is performed successfully (Shehab et al. 2018a).

The proposed hybrid algorithm, namely, CSBA, combines the optimization capabilities of MCSA and BA. This hybrid algorithm can overcome the MCSA limitations by exploiting the BA, which directly utilizes the most suitable global solution of the population to determine new positions for particles at each iteration (i.e., sharing information among solutions) (Shehab and Khader 2018). The results are expected to show that the proposed CSBA has fast convergence, high computational precision, and improved effectiveness in searching the optimal objective function value. The first performances of CSBA are carefully evaluated on 13 benchmark functions selected from the literature. The results indicate the effectiveness of the proposed algorithm compared with that of other methods for most benchmark cases. CSBA is then applied to synthetic data to extract the maxima of ODF, namely, CSBA-ODF. Section 7.3.2 describes the CSBA and CSBA-ODF.

7.3.1 Bat Algorithm

The BA introduced by Yang (2010b) emulates the echolocation behavior of bats. Many kinds of bats exist in nature. All of these kinds have similar behavior when navigating and hunting; however, they are different in terms of size and weight. Microbats extensively use the echolocation feature, which assists them in seeking for prey and/or avoid obstacles in a complete darkness. The behavior of microbats can be formulated by a novel optimization technique. This behavior is mathematically modeled as follows:

In BA, an artificial bat has position, velocity, and frequency vectors, which are updated during the course of iterations. BA can explore the search space through position and velocity vectors (or updated position vectors).

Each bat has a position X_i, frequency F_i, and Velocity V_i in a d-dimensional search space. The velocity, position, and frequency vectors are updated in Eqs. 7.1, 7.2, and 7.3.

$$V_i(t+1) = V_i(t) + (X_i(t) - Gbest) \times F_i \tag{7.1}$$

$$X_i(t+1) = X_i(t) + V_i(t+1) \tag{7.2}$$

where Gbest is the best solution obtained so far and F_i represents the frequency of the $i - th$ bat which is updated in each course of iteration as follows:

$$F_i = F_{min} + (F_{max} - F_{min}) \times \beta \tag{7.3}$$

where β is a random number of uniform distribution in [0,1]. The BA employed a random walk to improve its capability in exploitation as given below.

$$x_{new} = x_{old} + \varepsilon A_t \tag{7.4}$$

where ε is a random number in [−1,1], and A is the loudness of emitted sound. At each iteration, the loudness and pulse emission (r) are updated as follows:

$$A_i(t+1) = \alpha A_i(t) \tag{7.5}$$

$$r_i(t+1) = r_i(0)(1 - e^{(-\gamma \times t)}) \tag{7.6}$$

where α and γ are the constant parameters that lie between 0 and 1 and are used to update loudness rate A_i and pulse rate (r_i). The loudness and pulse emission rates essentially provide a mechanism for automatic control and auto-zooming into the region with promising solutions. The pseudo-code of the algorithm is presented in Algorithm 6.

Algorithm 6 Basic bat-inspired algorithm

```
Initialize the bat population X_i (i = 1, 2, .., n) and V_i
Define pulse frequency F_i
Initialize pulse rate r_i and the loudness A_i
while t < Maxnumberofiterations do
   Generate new solutions by adjusting frequency,
   Updating velocities and positions [Eqs. (7.1) to (7.1)]
   if rand > r_i then
      Select a solution among the best solutions randomly
      Generate a local solution around the selected best solution
   end if
   Generate a new solution by flying randomly
   if rand < A_i and f(X_i) < f(x*) then
      Accept the new solutions
      Increase r_i and reduce A_i
   end if
   Rank the bats and find the current Gbest
end while
```

7.3.2 CSBA and CSAB-ODF

On the basis of introducing MCSA and BA in previous sections, this section contains the details of merging MCSA with the components of BA. This merger avoids the solutions with poor fitness to increase the quality of solutions and focuses on exploitation to improve slow convergence, which leads to improving search efficiency. The following paragraphs present further details of the CSBA, followed by the details of CSBA-ODF.

In general, *Lévy* flight is applied to optimization and optimal search and proves its efficiency by achieving favorable results, which signify a promising start (Pavlyukevich 2007; Yang and Deb 2013; Abualigah et al. 2018b). Therefore, MCSA is balanced between exploration and exploitation (Layeb 2011; Li et al. 2014). However, solutions are poorly exploited at times, such as in cases when large steps or moves generate a new solution that is either far from the last solution or out of the boundary (Shehab et al. 2019b). By contrast, the effect will be unnoticeable when the step is small. CSBA is proposed to overcome this drawback of the MCSA by utilizing the advantages of BA, which can provide fast convergence at the initial stage by switching from exploration to exploitation (Yang and He 2013; Abualigah et al. 2018d). CSBA is used to increase the quality of the solutions, enhance performance, and avoid being trapped in the local optima (Yılmaz and Küçüksille 2015).

Figure 7.1 illustrates the CSBA flowchart, which is divided into three parts. The first part is initialization and comparison between the solutions of *Lévy* flight and tournament selection to send it to the second part. The second part is surrounded by a red stripe, which is the BA component. A new solution is generated in this part on the basis of solution i from the first part, and pulse frequency F_i, loudness A_i, and pulse rate r_i are defined, as mentioned in Eqs. 7.3, 7.5, and 7.6, respectively. Velocities and positions (Eqs. 7.1 to 7.2) are updated, thereby allowing the search for all possible solutions around the best solution (i.e., increase in exploitation). A solution is then randomly selected among the best solutions, and a local solution is generated around the selected best solution after checking r_i (i.e., intensive local search). The new solution is selected by comparing pulse frequency with the current solution. The second part provides a high chance for the best solution to exceed the last condition in the third part, which is called probability $P\alpha$ (Sect. 3.3). Thus, the best solution of the CSBA should be efficient. The first and third parts belong to the MCSA, and the second part belongs to the BA.

The CSBA-ODF is proposed to overcome the drawbacks of the MCSA-ODF. This version of CSA-ODF contains MCSA-ODF and components of BA. Algorithm 7 presents the pseudo-code of the CSBA-ODF, which follows the same procedures that are mentioned in the CSBA (i.e., divided into three main parts). A difference is that one solution (i.e., the best solution between the solutions of *Lévy* flight and tournament) is selected in the CSBA, whereas three solutions (i.e., the best solution has the preference to continue the other solutions to guarantee that no problem [plan B]) are selected in CSBA-ODF. This difference can be observed in Line 11, wherein

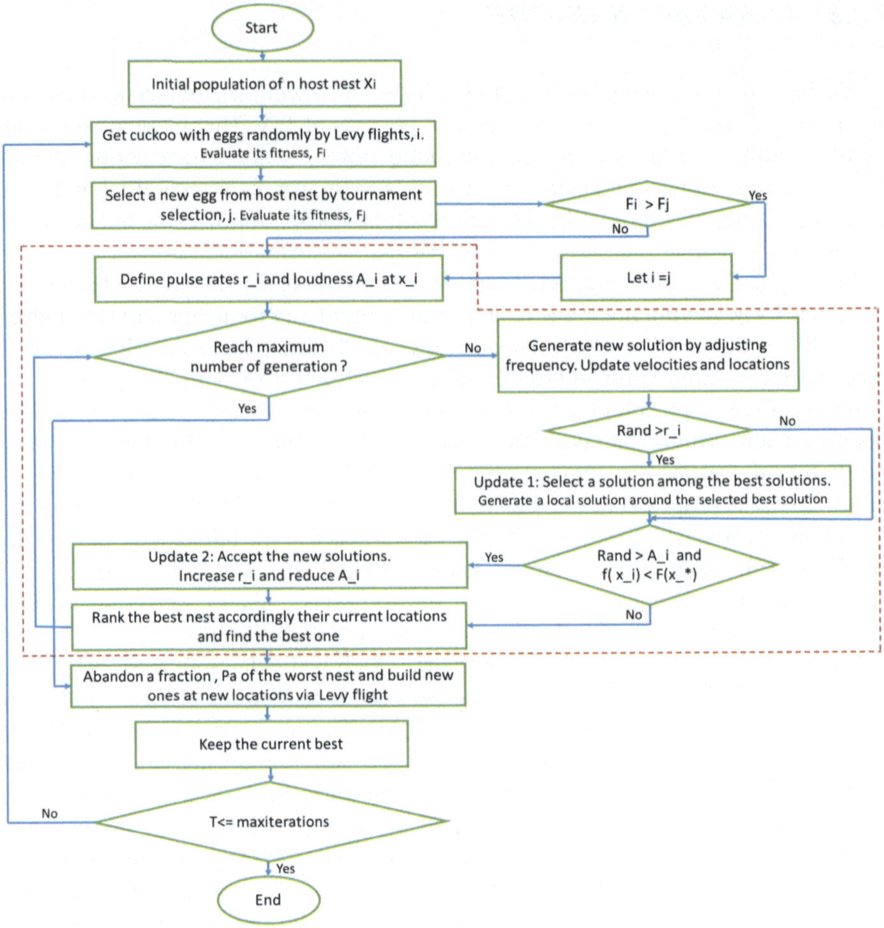

Fig. 7.1 Flowchart of CSBA

the while loop is started to determine the velocity and position for each solution of the three solutions to find the new solutions.

The CSBA-ODF mainly aims to minimize the angular resolution among the detecting maxima and prove its capability by handling different BCSA values (i.e., initial solution and step size) and its performance robustness by managing the different *b*-value and SNRs (i.e., increase image accuracy). The hybridization of MCSA-ODF and components of BA can achieve these goals given the advantages of MCSA-ODF and BA in the exploration search.

7.3.3 Experimental Results and Discussion

This section aims to study the performance of the proposed CSBA and CSBA-ODF, which is evaluated by performing two main experiments.

1. **benchmark optimization problems**:
   Thirteen well-known benchmark functions are used to test the CSBA performance, and their results are compared with the results of other algorithms (i.e., HS, KH, GA, BA, and MCSA). These benchmark functions are divided into unimodal and multimodal problems.
2. **Synthetic Data**:
   Synthetic data are similar to the data used in Sects. 4.4.3 and 6.5.

Algorithm 7 CSAHC-ODF

Objective function $f(X)$
while $t < Max\_iterations$ **do**
Generate initial population includes SH coordinates $xi_{(i=1,2,...,u)}$, where $x_i = (\theta_i, \phi_i)$
while stopping criterion **do**
Get a solution by *Lévy* flights: Max_i /** assume that the best solution**/
Check if the Max_i in the search space
Evaluate its quality/ fitness $F(Max_i)$ by Eq.4.5
Call the tournament selection (Algorithm 3) to determine two solutions Max_j ***and*** Max_k /** neighbors of the Max_i **/
Check if the solutions Max_j and Max_k are in the search space
Evaluate its quality/ fitness $F(Max_j)$ and $F(Max_k)$ by Eq.4.5 /** send the three solution to the second step for more expolitation**/
while (s<=3) **do**
Define pulse frequency F_s
Initialize pulse rate r_s and the loudness A_s
while $t < Maxnumberofiterations$ **do**
Generate new solutions by adjusting frequency,
Updating velocities and positions [equations (7.1) to (7.1)]
if $rand > r_s$ **then**
Select a solution among the best solutions randomly
Generate a local solution around the selected best solution
end if
Generate a new solution by flying randomly
if $rand < A_s$ and $f(X_s) < f(x^*)$ **then**
Accept the new solutions
Increase r_s and reduce A_s
end if
Rank the solution and find the best solution B_s
s= s+1
end while
end while
Get the Minimum and Maximum between B_1, B_2, and B_3 : MIN and MAX /** If the MIN pass from the last checking Pα the other will pass **/
Check the MIN by Probability $(P\alpha)$ /** if the MIN discover: will be abandoned the solutions and build new solution by *Lévy* flights **/
end while
Maximum= Max
Store the other two solutions (precaution)
Labeling the points that constitute the vicinity of the detected *Maximum* and its antipodal
end while
Postprocess results and visualization

7.3.3.1 Experimental Design

The quality of the CSBA method is proven by a set of global numerical optimization problems through various experiments performed in benchmark functions, which are implemented in the same conditions as in (Wang et al. 2016b). In the present work, 13 benchmark functions are used to analyze the CSBA method. Table 4.2 summarizes the expressions of the 13 benchmark functions. The information of all benchmark functions can also be found in (Jamil and Yang 2013; Shehab et al. 2016).

CSBA is initially compared with the global optimization problems of five optimization algorithms, namely, KH, HS, GA, BA, and MCSA. In the experiments, similar parameters for BA and CSBA are set with $P\alpha = 0.2$, loudness $A = 0.5$, and pulse rate $r = 0.5$. The parameters used for KH, HS, and GA are the same parameters used in each original algorithm. For other parameters, population size, frequency minimum, frequency maximum, and maximum generation are set to 20, 0, 2, and 100,000, respectively. A total of 100 implementations are performed for each algorithm on each benchmark function to decrease the influence of randomness. Tables 7.1 and 7.2 list the mean and best normalization of the optimization results of the six methods.

CSBA-ODF is applied to synthetic data after proving the efficiency of the CSBA through benchmark functions, as presented in Sect. 7.3.3.2.

Table 7.1 Mean normalised optimization results

	KH	HS	GA	BA	MCSA	CSBA
F1	1.37E+03	1.23E+03	1.59E+02	8.35E+03	3.49E+02	**1.00**
F2	3.42E+03	6.01E+04	2.38E+02	3.41E+03	1.22E+00	**1.00**
F3	**1.00**	1.97E+00	**1.00**	1.31E+00	**1.00**	**1.00**
F4	2.02E+02	1.90E+02	2.11E+01	7.52E+01	2.01E+01	**1.00**
F5	6.99E+02	2.24E+02	**1.00**	4.74E+02	2.09E+01	1.34E+00
F6	6.97E+04	6.52E+05	8.22E+02	8.40E+04	4.17E+02	**1.00**
F7	3.73E+02	7.80E+02	**1.00**	1.20E+02	8.16E+01	1.51E+00
F8	1.52E+03	1.35E+04	8.90E+02	3.80E+03	**1.00**	**1.00**
F9	3.97E+02	1.79E+03	6.16E+01	1.72E+02	1.28E+01	**1.00**
F10	4.01E+00	2.73E+02	1.02E+00	1.03E+01	**1.00**	**1.00**
F11	6.74E+03	5.32E+03	1.04E+02	8.95E+03	4.50E+02	**1.00**
F12	1.09E+01	1.31E+02	7.97E+00	3.45E+02	**1.00**	**1.00**
F13	4.50E+02	4.42E+03	4.50E+01	**1.00**	1.01E+00	**1.00**
Total[a]	**1**	**0**	**3**	**1**	**4**	**11**

[a]The total (last row) refers to the summation of all number 1.00, which indicates the minimum normalization value (i.e., when the total is higher, the performance is better)

Table 7.2 Best normalised optimization results

	KH	HS	GA	BA	MCSA	CSBA
F1	2.19E+04	1.30E+05	1.25E+03	6.66E+04	2.13E+03	**1.00**
F2	2.10E+09	1.18E+06	7.63E+03	5.31E+05	1.48E+04	**1.00**
F3	1.32E+03	1.26E+04	1.26E+02	1.04E+03	**1.00**	**1.00**
F4	5.14E+03	1.10E+04	8.30E+02	6.75E+03	**1.00**	**1.00**
F5	1.30E+02	1.67E+03	3.33E+01	1.95E+02	1.01E+02	**1.00**
F6	1.38E+06	1.00E+07	**1.00**	9.46E+06	4.48E+01	**1.00**
F7	2.61E+05	1.74E+06	**1.00**	8.58E+03	2.61E+03	2.39E+00
F8	5.87E+04	5.79E+06	1.48E+02	1.94E+04	1.17E+01	**1.00**
F9	2.82E+02	9.74E+02	3.85E+00	1.83E+02	9.10E+01	**1.00**
F10	1.20E+03	8.57E+04	5.12E+01	1.81E+02	2.65E+01	**1.00**
F11	6.62E+04	2.17E+05	4.35E+01	1.10E+02	3.12E+02	**1.00**
F12	3.14E+06	7.58E+06	9.77E+02	7.95E+04	7.68E+02	**1.00**
F13	4.50E+02	4.36E+03	4.50E+00	5.78E+03	1.26E+02	**1.00**
Total[a]	**0**	**0**	**2**	**0**	**2**	**12**

[a]The total (last row) refers to the summation of all number 1.00, which indicates the minimum normalization value (i.e., when the total is higher, the performance is better)

7.3.3.2 Experimental Results

1. **Benchmark Function**
   Table 7.1 illustrates the mean normalization of the optimization results. The CSBA method is the most effective in determining the minimum objective function on 11 out of the 13 benchmarks, which are F1-F4, F6, and F8-F13 (i.e., all benchmark functions, except F5 and F7). As mentioned previously (Sect. 6.5.2), F5 and F7 belong to the unimodal functions. The MCSA achieves favorable results (nearly the best result: F5 = 1.34E + 00 and F7 = 1.51E + 00). Therefore, the CSBA performance is improved compared with the MCSA performance that still requires further improvements to achieve the best results. BCSA ranks second and performs best on F3, F8, F10, and F12, followed by GA, which performs best on F3, F5, and F7. HS performs best on F3 and BA on F13.
   Table 7.2 indicates that the CSBA performs best on 12 out of the 13 benchmarks, (i.e., except F7). BCSA and GA are the second most effective, performing best on benchmarks F10, F13 and F4, and F6, respectively.
   The most representative convergent curves are illustrated in Fig. 7.2a–h. The values in the figures are the mean function optima, which are the true values.
   Figure 7.2a presents that CSBA can find better solutions than those of all other methods. HS converges sharply at the initial search stage; however, the global minimum slightly decreases when the HS is trapped into the sub-minima. BA is near the CSBA in the initial stage, but the difference increases in the second

stage. BA, BCSA, GA, and KH initially move to the best solutions, whereas GA converges to the better minimum later than the other methods. CSBA is the best of all methods.

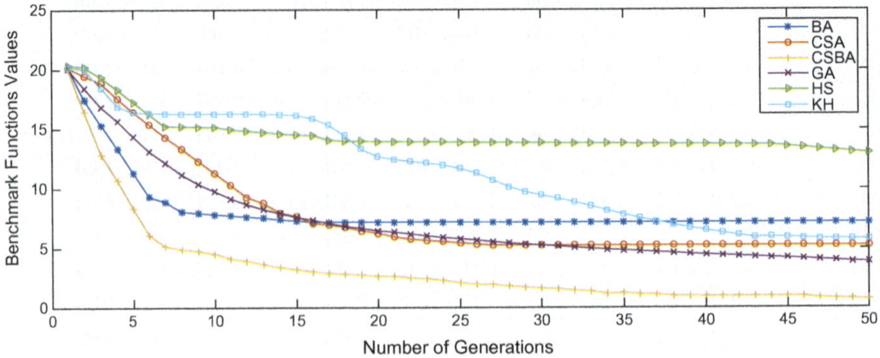

(a) Ackley Function

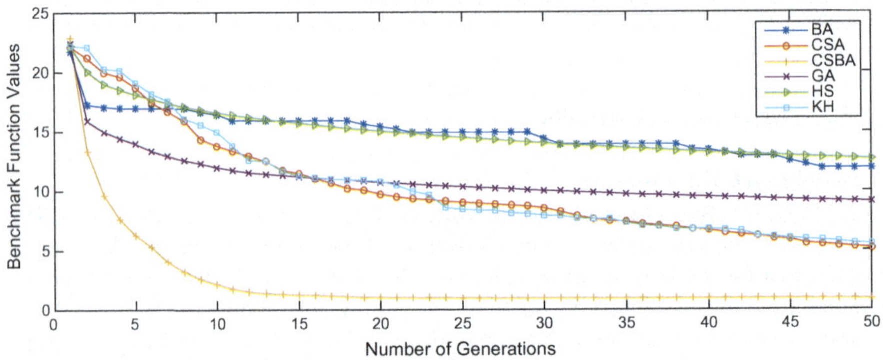

(b) Penalty #1 Function

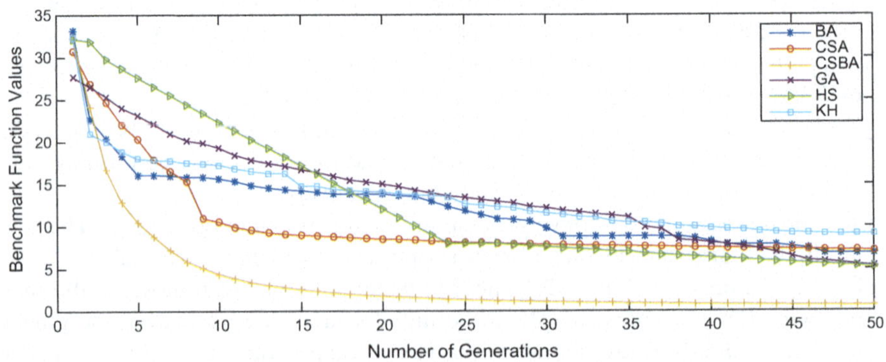

(c) Penalty #2 Function

Fig. 7.2 Performance comparison for the benchmark functions

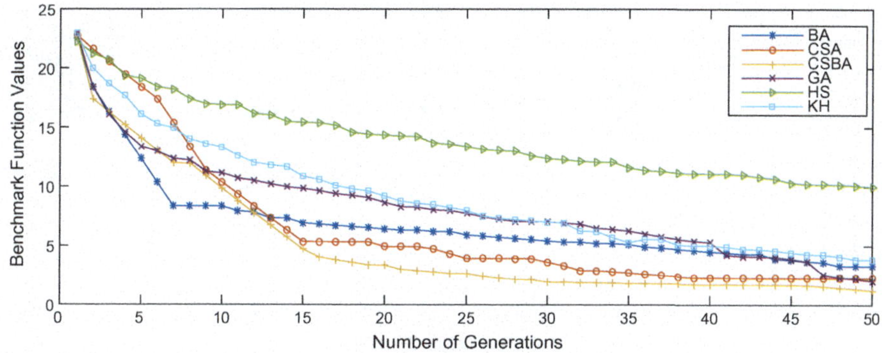

(d) Rastrigin Function

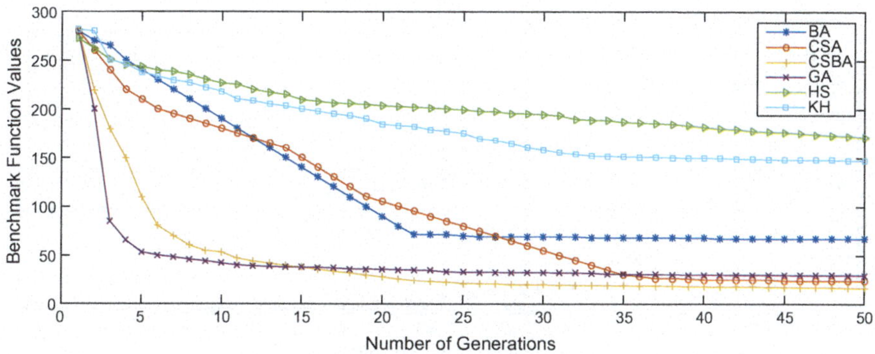

(e) Rosenbrock Function

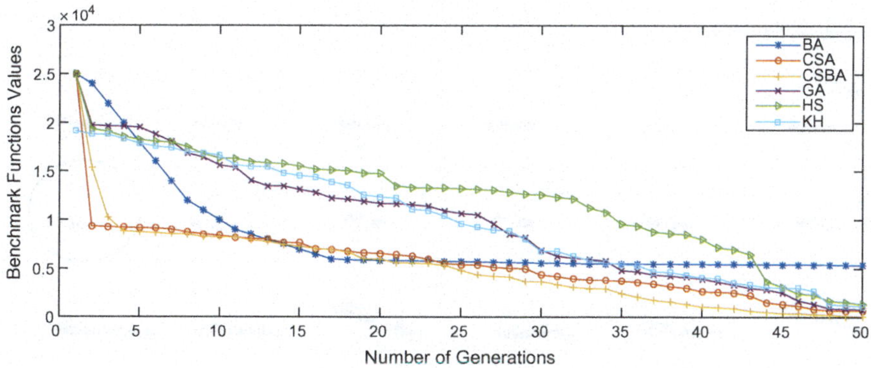

(f) Schwefel 1.2 Function

Fig. 7.2 (continued)

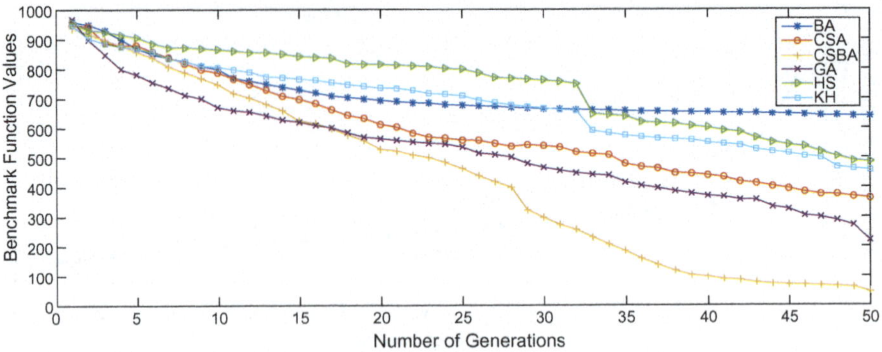

(g) Schwefel 2.22 Function

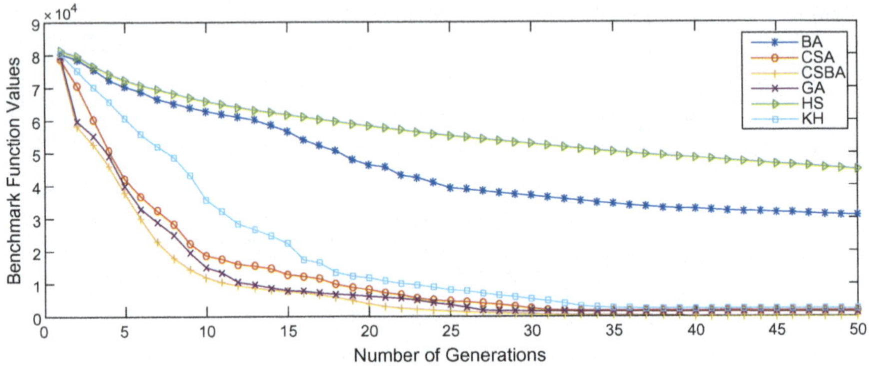

(h) Sphere Function

Fig. 7.2 (continued)

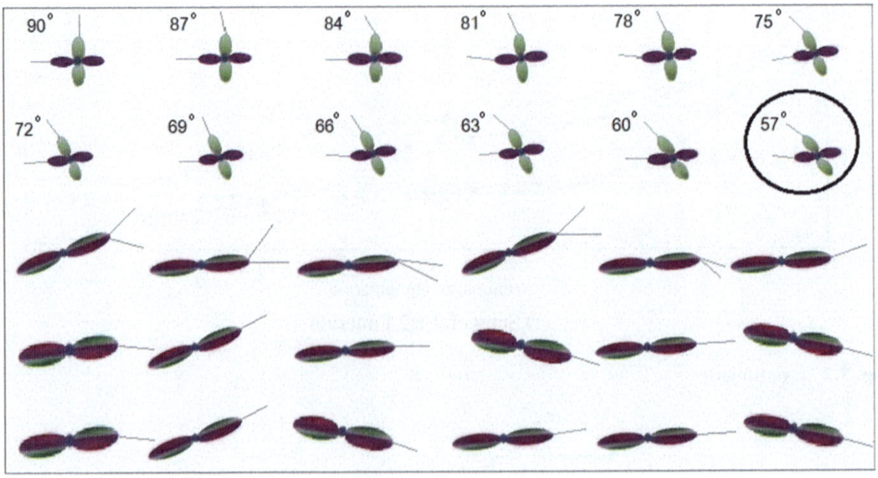

Fig. 7.3 The ability of the CSBA-ODF to detect maxima separated by 57°

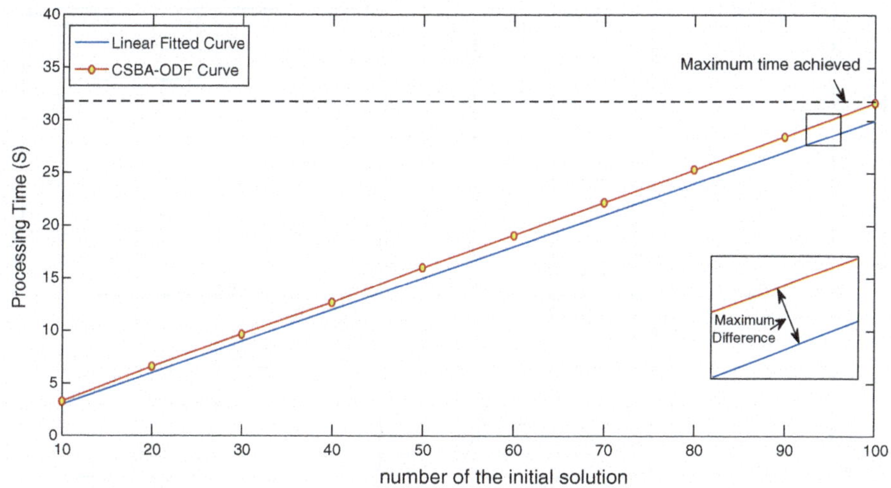

Fig. 7.4 Influence of the number of initial solutions on the processing time of the CSBA-ODF

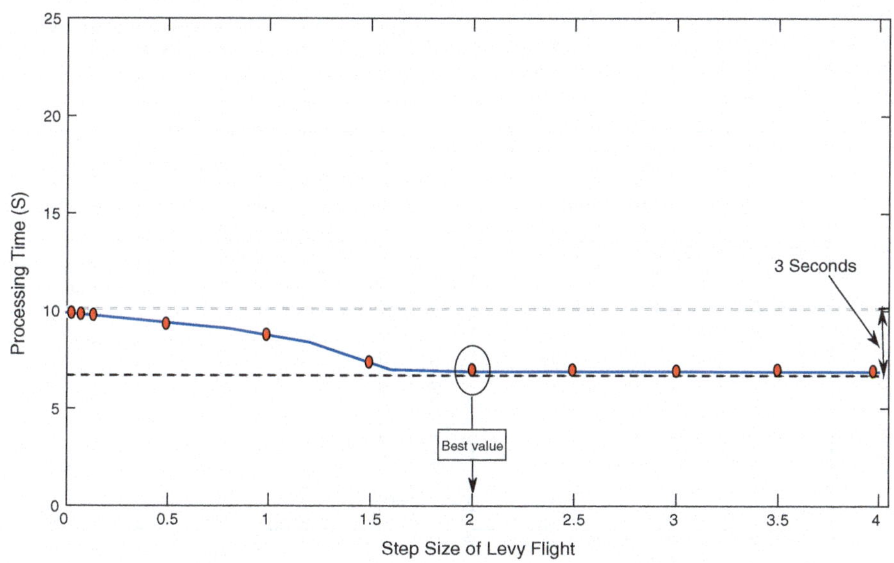

Fig. 7.5 Influence of the step size of *Lévy* Flights on the processing time of the CSBA-ODF

Figure 7.2b illustrates that CSBA is the fastest method for finding the best solution in the first part, and GA performs the second best solution; BCSA and KH are the best performers in the second part. All algorithms begin the optimization process at the beginning of nearly the same point, whereas CSBA has a more stable convergent speed than the other methods. Difficulties are encountered in

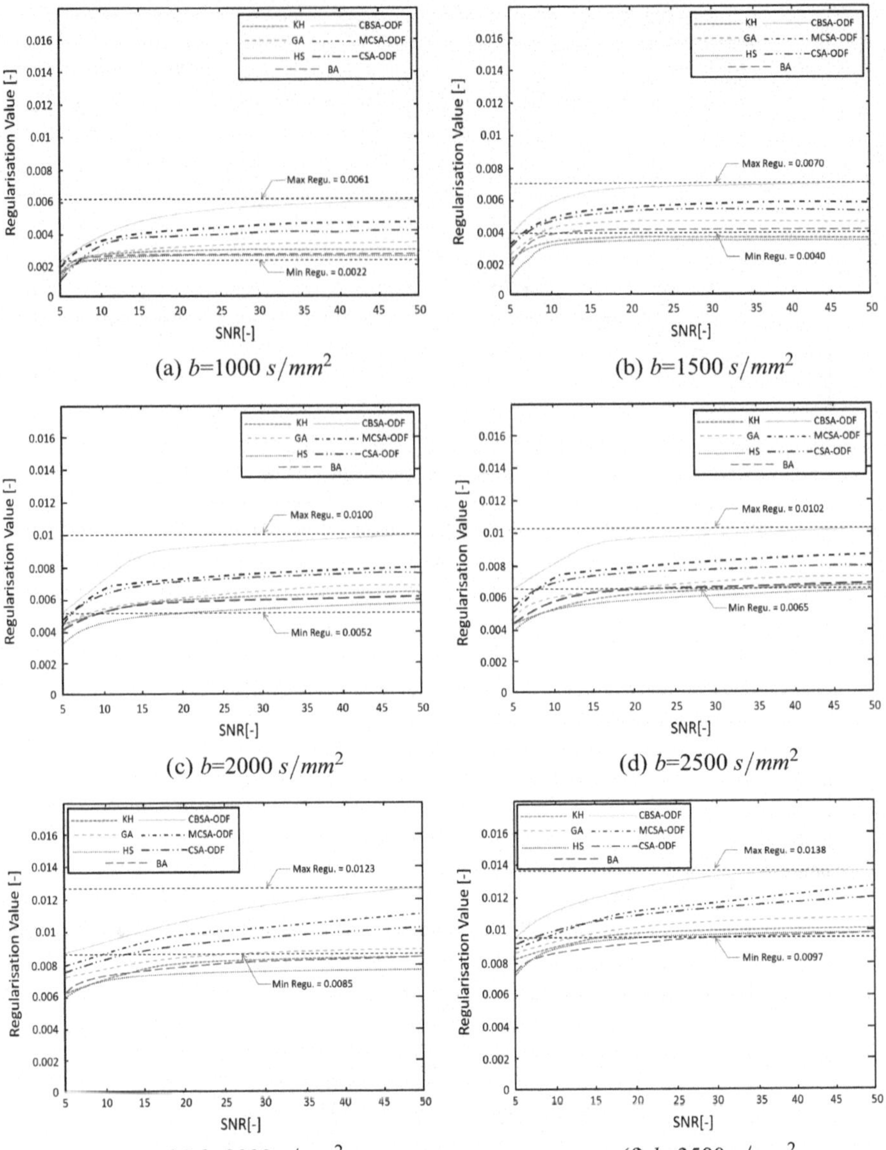

(a) b=1000 s/mm^2

(b) b=1500 s/mm^2

(c) b=2000 s/mm^2

(d) b=2500 s/mm^2

(e) b=3000 s/mm^2

(f) b=3500 s/mm^2

Fig. 7.6 The influence of the b-value on the HS, KH, BA, GA, CSA-ODF, MCSA-ODF, and CSBA-ODF

finding the best results in the original BCSA compared with the other methods. In this case, CSBA is the most efficient and fastest method in finding the best global function values among the six methods.
Figure 7.2c depicts that the results are nearly similar to the results in Fig. 7.2b, which implies that the original BCSA encounters several problems in determining the best global solution compared with the other methods. The results in the initial stage display that KH achieves better results than those of CSBA. However, the final results illustrate that CSBA obtains the best results, whereas KH generates the worst. HS has a faster convergent speed than the other methods and achieves the closest results to CSBA.
Figure 7.2d presents that the CSBA performs equally with all the methods. BA achieves the best results at the 11th generation. CSBA converges in a stable state in this case. The combination of BCSA and BA demonstrates the generation of a favorable performance.
Figure 7.2e depicts that CSBA has the best performance among the six methods, and BCSA ranks second. GA has the third best performance with a relatively slow and stable convergence rate. Similarly, the results for all the methods are near one another with CSBA and a slight advantage, as demonstrated in Fig. 7.2f.
Figure 7.2g presents that CSBA in the unimodal case performs better than the other methods, especially in the second part of the results. GA outperforms CSBA until the 15th generation and BCSA until the end.
Figure 7.2h shows that CSBA has the top performance among the six methods in the optimization process, followed by GA, BCSA, and KH.
The analysis of Figs. 7.2a–h reveals that our proposed metaheuristic CSBA method significantly outperforms the other methods.

2. **Synthetic Data**

   - **Angular resolution of the CSBA-ODF**
     In this section, we aim to determine the angular resolution of the ODF as the smallest crossing angle for which two distinct peaks are detectable.
     Figure 7.3 exhibits the detected ODF maxima that start from the upper left with a separation angle of 90° among its maxima. We decrease 3° at each time by moving from left to right in each line. CSBA- ODF proves its capability to detect the ODF maxima for separation angles equal and above 57°. This result is significant compared with those of the CSA-ODF and MCSA-ODF, which detects the ODF maxima for separation angles 66° and 63°, respectively (Figs. 5.1 and 6.3).
   - **Influence of control parameter**

   (a) **Influence of the number of the initial solutions (initial nests)**
   This portion follows Sect. 5.4.2(c), which presents the effect of the initial solution of nests. Different values of initial solutions are used to evaluate the CSBA-ODF performance. The average of 1000 run times with a *b*-value of 3000 sec/mm^2 is computed in the experiments. Figure 7.4 illustrates the CSBA-ODF results from the beginning until the initial solution = 40 are near the fitted linear. However, the gap between the two lines in the second part

(i.e., CSBA-ODF and fitted linear) increases until the difference between the two lines is approximately 1.7 s, the CSBA-ODF's result is better than the MCSA-ODF's result which it was 2.3 s. However, the CSBA-ODF still requires more improvements to achieve favorable results in large values of the initial solutions.

(b) **Influence of the *Lévy* Flights**
This version is characterized by its control of the search way, which can perform deep (local search) and distant searches (global search). Therefore, CSBA is not influenced by a change in values of the step size of a *Lévy* flight (the values are mentioned in Sect. 5.4.2(c)). The results in Fig. 7.5 indicate that the effect of the step size on time equal 3 s (the time difference between the lowest and highest values of the step size). BA can control step size by maintaining the processes within the search space (i.e., it uses the best solution for each generation in the next generation, thus alleviating the effect of step size). The effects of step size on the CSA-ODF and MCSA-ODF are 7.5 and 5.5 s, respectively (Figs. 5.2 and 6.5). Finally, it can notice that the best solution achieved with the minimum processing time when the step size 2.

- **Influence of SNR and *b*-value**:
In this experiment, HS, KH, BA, GA, CSA-ODF, MCSA-ODF, and CSBA-ODF are compared by using different *b*-value (1000, 1500, 2000, 2500, 3000, and 3500), different SNRs (5, 10, 15, 20, 25, 30, 30, 35, 40, 45, and 50), and average of 100 run times for each value. The experiments mentioned in Sect. 5.4.2(c) are divided into groups (i.e., subfigures), in which each group shows the results of the effects of SNR and *b*-value to evaluate the robustness and performance of each algorithm by determining the value of the regularization parameter, which is responsible for the exterior shape of the ODF. Each *b*-value is fixed for all SNR values.
Figure 7.6 illustrates the influences of SNR and *b*-value on the HS, KH, BA, GA, CSA-ODF, MCSA-ODF, and CSBA-ODF to determine the most accurate method to be used for real data. The regularization value increases gradually by increasing each of the *b*-value and SNR. However, the regularization value increases slightly when SNR = 30 and above with low *b*-values (i.e., *b*-value = 1000, 1500, 2000, 2500), as depicted in Fig. 7.6a–d, thereby indicating the limitations of SNR and non-limited *b*-value influences. All results demonstrate that the CSBA-ODF achieves the highest regularization, which means high accuracy, thereby leading to decreased regularization values (i.e., the smoothness of the ODF shape) with increased SNRs. The performances of all algorithms are close to one another, which outperform the MCSA-ODF performance by maintaining the highest regularization value.

7.4 Hybridization of CSBA with Hill Climbing Algorithm

A new hybrid optimization approach, namely, CSAHC is developed in this section by hybridizing the CSBA with HC. Then, the CSAHC is applied to a synthetic dataset and is called CSAHC-ODF. The CSAHC aims to improve the performances of CSBA and CSBA-ODF in the local search. The performances can be evaluated via two ways: (i) CSAHC: uses 13 benchmark functions and compares the results with those of KH, HS, GA, BA, and CSBA to guarantee its effectiveness; and (ii) CSAHC-ODF: uses synthetic data to check the CSAHC performance in solving ODF problem.

The structure of this section is organized as follows. Section 7.4.1 briefly describes HC. Section 7.4.2 presents the CSAHC and CSAHC-ODF approaches. Section 7.4.3 indicates the experiments and discussions.

7.4.1 Hill Climbing Algorithm

HC is a mathematical optimization technique that belongs to the family of local search (Schaerf and Meisels 1999). This technique searches for an appropriate solution in the neighborhood by evaluating the current state. If the solution is a goal state, then HC returns to it and quits. Otherwise, the current state is updated. HC loops until a solution is found or until no new operators remain to be applied in the current state. Two steps are performed inside the loop. The first step is selecting an operator that has not yet been applied to the current state and applying it to produce the new state. The second step is evaluating the new state. Figure 7.7 displays the HC flowchart, which proves the simplicity of HC Abualigah et al. (2017b).

The main objective on the basis of the working principle of HC is to continue searching for a better solution than that of the current solution. Thus, HC improves the quality of a solution most of the time (Burke and Newall 2002; Shehab et al. 2019a). The stopping criteria of HC are responsible for stopping the search without any change if the HC cannot determine a favorable neighbor.

HC has several advantages, such as the capability to be easily adjusted to the problem at hand. Nearly any aspect of the algorithm may be changed and customized. For example, HC can be used in conversions and discrete domains (Alajmi et al. 2011; Rubio and Gámez 2011).

7.4.2 CSAHC and CSAHC-ODF

CSBA proves its capability in exploitation search by achieving better results than those of the other algorithms. However, CSBA undergoes the premature convergence problem for high-dimensional problems because the algorithm converges rapidly

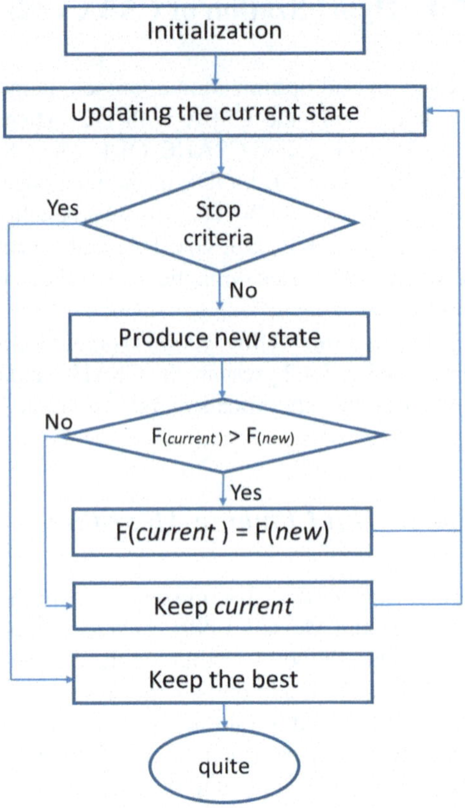

Fig. 7.7 Flowchart of the HC

Shehab et al. (2018b). CSAHC is proposed to solve this issue by hybridizing optimization algorithm, which is a combination of CSBA and HC. The CSAHC discovers many local optimum traps by using local and global searches, although the local search method is trapped at the local minimum point Shehab et al. (2017b).

Figure 7.8 shows the flowchart of the CSAHC. The flowchart is divided into three main parts as follows:

- The first part belongs to the CSBA, which starts with initial solutions and then selects one solution by using the tournament selection scheme. Velocity and position are added to the selected solution, and this step refers to the beginning of exploitation. The new solution is selected on the basis of the best solution in each generation.
- The second part can be observed in the middle of the flowchart (surrounded by a stripe), which includes HC. In this part, the work starts from the solution achieved by the first part. The solution neighbors are searched, and their fitness is compared.
- The best solution (from the second part) is moved to the third part, which includes the last check called Pα. The results should have high quality and accuracy after completing the three parts.

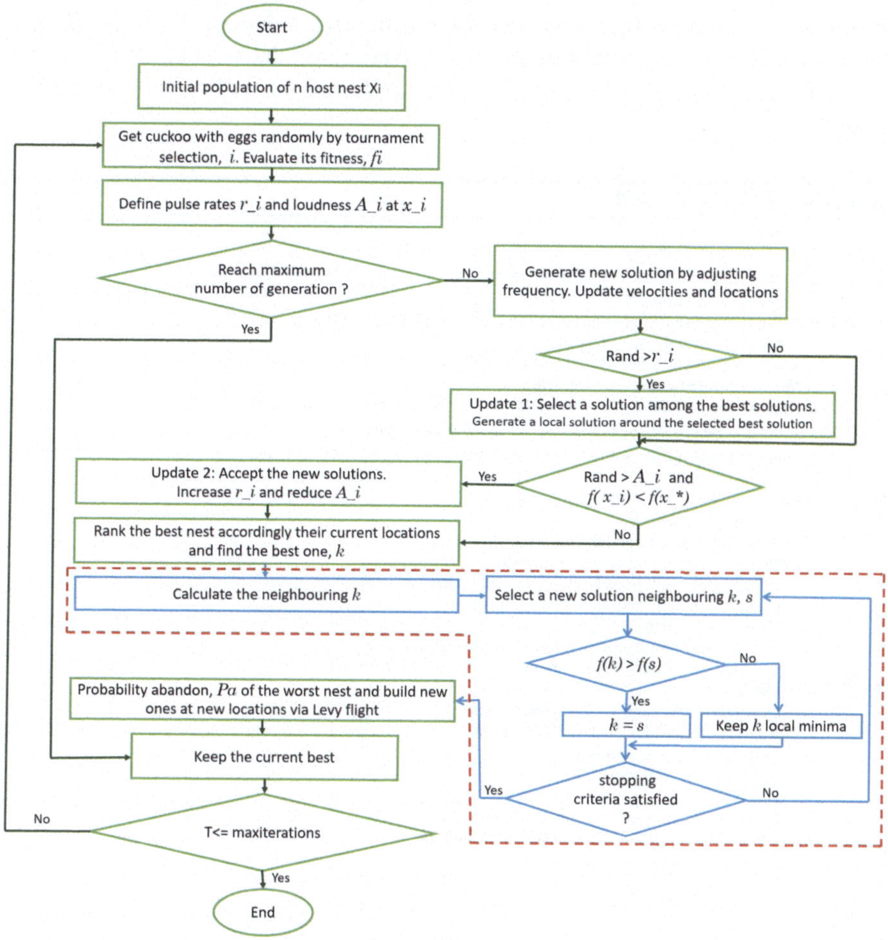

Fig. 7.8 Flowchart of the CSAHC Algorithm Shehab et al. (2018a)

CSAHC-ODF consists of adopting CSAHC to solve the ODF problem (i.e., extracting the ODF maxima). CSAHC-ODF is proposed to improve the previous version, which is the CSBA-ODF, by enhancing its results, such as angular resolution (i.e., to achieve a minimal separation angle), influences of the BCSA parameters (i.e., to increase its capability with the change in parameter values), and influences of SNR and b-value (i.e., to increase the regularization value for enhancing image quality). Algorithm 8 illustrates the pseudo-code of the CSAHC-ODF. This CSA-ODF version has nearly the same procedures as those of the CSBA-ODF, but the CSAHC-ODF adds a new step of local search through HC. This step starts in Line 27, where the function Explore($N(B_s)$) navigates the neighboring solutions $N(B_s)$ of B_s and moves to the first neighboring solution $B_{s(n)} \in N(B_s)$, which has a large fitness value (i.e., $F(B_{s(n)} > F(B_s)))$. This process is repeated until no further improvement is obtained.

Algorithm 8 CSAHC-ODF

Objective function $f(X)$
while $t < Max\_iterations$ **do**
Generate initial population includes SH coordinates $xi_{(i=1,2,....,u)}$, where $x_i = (\theta_i, \phi_i)$
while stopping criterion **do**
Get a solution by *Lévy* flights: Max_i /** assume that the best solution**/
Check if the Max_i in the search space
Evaluate its quality/ fitness $F(Max_i)$ by Eq.4.5
Call the tournament selection (Algorithm 3)
to determine two solutions Max_j ***and*** Max_k /** neighbors of the Max_i **/
Check if the solutions Max_j and Max_k are in the search space
Evaluate its quality/ fitness $F(Max_j)$ and $F(Max_k)$ by Eq.4.5 /** send the three solution to the second step for more exploitation**/
while (s<=3) **do**
Define pulse frequency F_s
Initialize pulse rate r_s and the loudness A_s
while $t < Maxnumberofiterations$ **do**
Generate new solutions by adjusting frequency,
Updating velocities and positions [equations (7.1) to (7.1)]
if $rand > r_s$ **then**
Select a solution among the best solutions randomly
Generate a local solution around the selected best solution
end if
Generate a new solution by flying randomly
if $rand < A_s$ and $f(X_s) < F(x^*)$ **then**
Accept the new solutions
Increase r_s and reduce A_s
end if
Rank the solution and find the best solution B_s
while (Local optimal solution is not reached) **do**
$B_{s(n)} = Explore(N(B_s))$
if $F(B_{s(n)} > F(B_s))$ **then**
$(B_s)= B_{s(n)}$
end if
end while
s= s+1
end while
end while
Get the Minimum and Maximum between B_1, B_2, and B_3: MIN and MAX /** If the MIN pass from the last checking Pα the other will pass **/
Check the MIN by Probability ($P\alpha$) /** if the MIN discover: will be abandoned the solutions and build new solution by *Lévy* flights **/
end while
Maximum= Max
Store the other two solutions (precaution)
Labeling the points that constitute the vicinity of the detected *Maximum* and its antipodal
end while
Postprocess results and visualization

7.4.3 *Experimental Results and Discussion*

This section indicates the study of the performance of the proposed CSAHC. The evaluated performance of CSAHC includes two main experiments.

1. **Benchmark Optimization Problems**:
   Thirteen well-known benchmark functions are used for comparison to test the CSAHC performance. Table 4.2 describes the benchmark functions. These benchmark functions are divided into unimodal and multimodal problems.
2. **Synthetic Data**:
   Synthetic data are similar to the data used in Sects. 4.4.3, 6.5, and 7.3.3.

7.4.3.1 Experimental Design

The quality of the CSAHC method is verified by a set of global numerical optimization problems through various experiments performed in benchmark functions, which are implemented in the same conditions as in (Wang et al. 2016b). In the present work, 13 benchmark functions are used to analyze the CSAHC method. The information of all benchmark functions can also be found in (Jamil and Yang 2013; Simon 2008).

In these experiments, the same values are used for the parameters, such as $n = 20$ and $P\alpha = 0.20$. All tests are run 100 times. The mean and best normalizations for each algorithm are demonstrated in Tables 7.3 and 7.4, respectively.

The CSAHC-ODF is applied to synthetic data after verifying the CSAHC efficiency, as mentioned in Sect. 7.4.3.2.

7.4.3.2 Experimental Results

1. **Benchmark Functions**
   Table 7.3 denotes that CSAHC performs the best mean normalization on 13 of the 13 benchmark functions. This result illustrates the performance efficiency of the CSAHC, which achieves the best solution of benchmark function F7 (i.e., it improves the CSBA, which does not achieve the best solution of F7). CSBA achieves the second position with 7 of the 13 benchmark functions, followed by GA, KH, BA, and HS, with 4, 2, 2, and 1 benchmark, respectively. Most algorithms achieve the best solutions in F4 and F13 due to the ease of their equations to find the optimal solutions.
   Table 7.4 summarizes the best normalization of the results. CSAHC is the most effective in determining the minimum objective function on 12 out of the 13 benchmarks, which are F1-F6 and F8-F13. CSAHC does not achieve the best normalization for benchmark function F7, which is called *Rosenbrock*; this function is located under the unimodal function, thereby indicating that it does not have many local optima with the number increasing exponentially with dimen-

Table 7.3 Mean normalised optimization results

	KH	HS	GA	BA	CSBA	CSAHC
F1	3.26E+05	4.29E+06	2.98E+02	8.22+E05	**1.00**	**1.00**
F2	2.51+E04	6.15E+04	9.32E+02	3.09E+04	**1.00**	**1.00**
F3	1.81E+03	5.08E+03	7.45E+01	6.64E+03	**1.00**	**1.00**
F4	**1.00**	5.44E+00	**1.00**	**1.00**	**1.00**	**1.00**
F5	3.25E+04	8.85E+04	3.63E+02	2.85E+04	5.11E+01	**1.00**
F6	4.36E+04	8.72E+05	2.21E+02	1.25E+04	8.29E+01	**1.00**
F7	1.27E+05	4.52E+05	6.78E+01	2.68E+05	5.02E+01	**1.00**
F8	4.02E+03	7.69E+04	9.33E+01	5.43E+03	**1.00**	**1.00**
F9	5.46E+02	7.63E+03	4.28E+01	7.09E+02	1.09E+01	**1.00**
F10	3.31E+02	1.0E+03	**1.00**	4.21E+02	3.02E+02	**1.00**
F11	7.25E+01	9.79E+02	**1.00**	8.59E+01	8.33E+01	**1.00**
F12	1.41E+03	5.93E+03	7.02E+01	2.36E+03	**1.00**	**1.00**
F13	**1.00**	**1.00**	**1.00**	**1.00**	**1.00**	**1.00**
Total[a]	**2**	**1**	**4**	**2**	**7**	**13**

[a]The total (last row) refers to the summation of all number 1.00, which indicates the minimum normalization value (i.e., when the total is higher, the performance is better)

Table 7.4 Best normalised optimization results

	KH	HS	GA	BA	CSBA	CSAHC
F1	6.89E+02	6.22E+03	2.72E+02	5.32E+03	**1.00**	**1.00**
F2	7.25E+03	2.36E+04	9.64E+01	8.87E+02	2.10E+01	**1.00**
F3	5.17E+02	2.02E+03	1.93E+01	2.41E+03	**1.00**	**1.00**
F4	5.67E+01	8.80E+02	2.11E+00	7.11E+02	**1.00**	**1.00**
F5	1.89E+04	1.65E+05	**1.00**	1.84E+04	2.10E+01	**1.00**
F6	9.78E+03	7.35E+04	5.01E+02	9.42E+03	5.75E+01	**1.00**
F7	2.74E+02	9.52E+03	9.85E+01	**1.00**	8.89E+00	1.01E+00
F8	7.35E+03	2.59E+04	9.15E+01	7.49E+03	1.75E+01	**1.00**
F9	1.28E+02	7.13E+02	3.46E+01	1.69E+02	7.07E+01	**1.00**
F10	7.30E+01	6.84E+02	4.18E+00	2.28E+02	5.18E+01	**1.00**
F11	3.82E+02	8.63E+03	6.49E+01	8.32E+03	4.26E+01	**1.00**
F12	1.68E+03	1.24E+04	8.55E+02	1.54E+03	4.73E+000	**1.00**
F13	**1.00**	1.08E+01	**1.00**	**1.00**	**1.00**	**1.00**
Total[a]	**1**	**0**	**2**	**2**	**4**	**12**

[a]The total (last row) refers to the summation of all number 1.00, which indicates the minimum normalization value (i.e., when the total is higher, the performance is better)

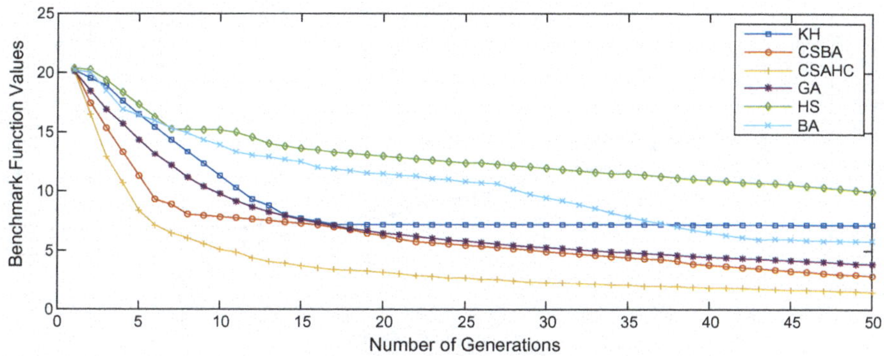

(a) Ackley Function

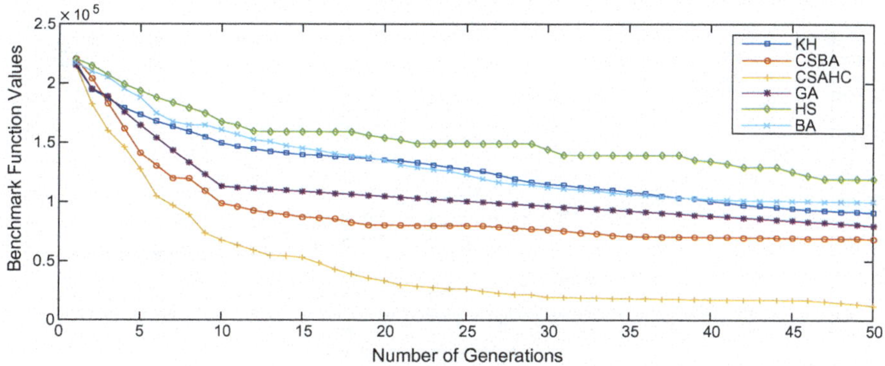

(b) Penalty #1 Function

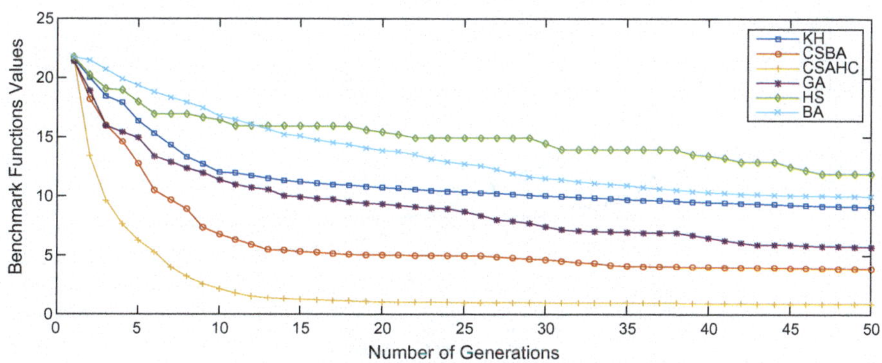

(c) Penalty #2 Function

Fig. 7.9 Performance comparison for the benchmark functions

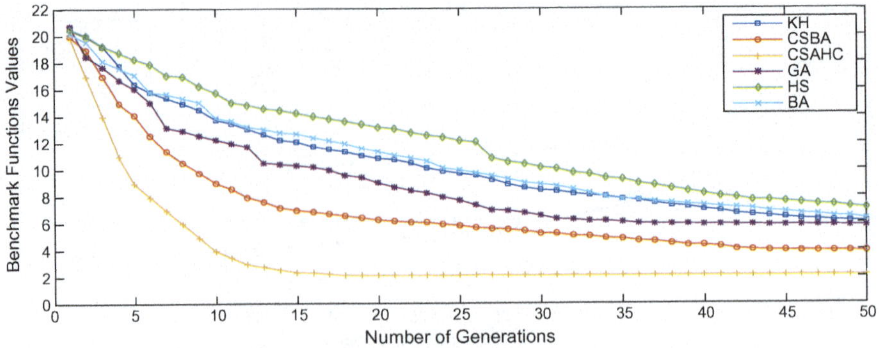

(d) Rastrigin Function

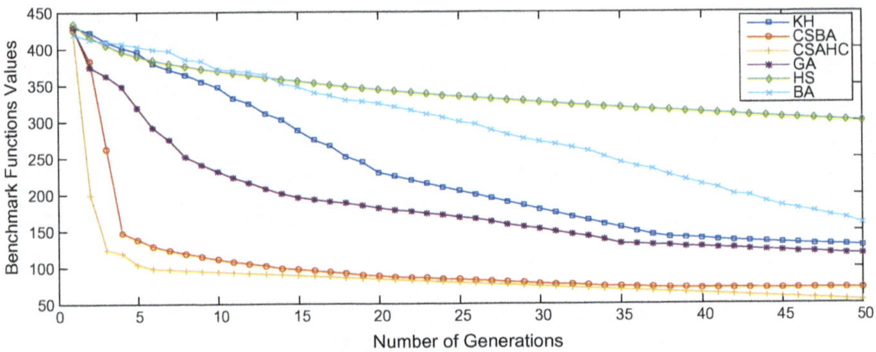

(e) Rosenbrock Function

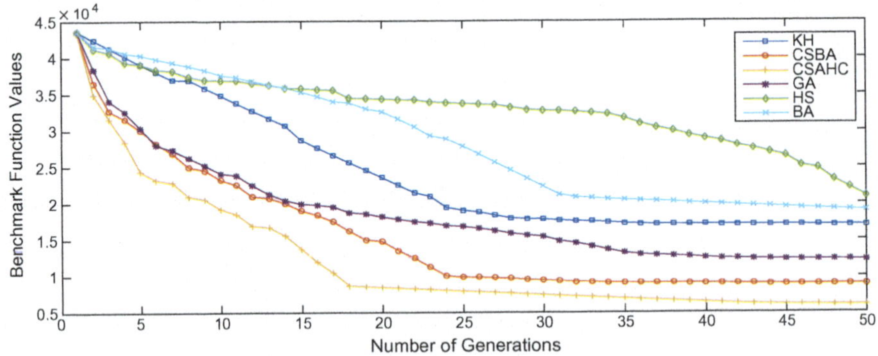

(f) Schwefel 1.2 Function

Fig. 7.9 (continued)

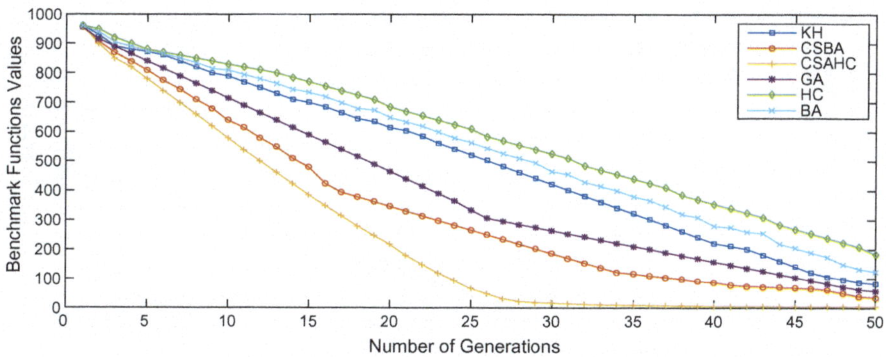

(g) Schwefel 2.22 Function

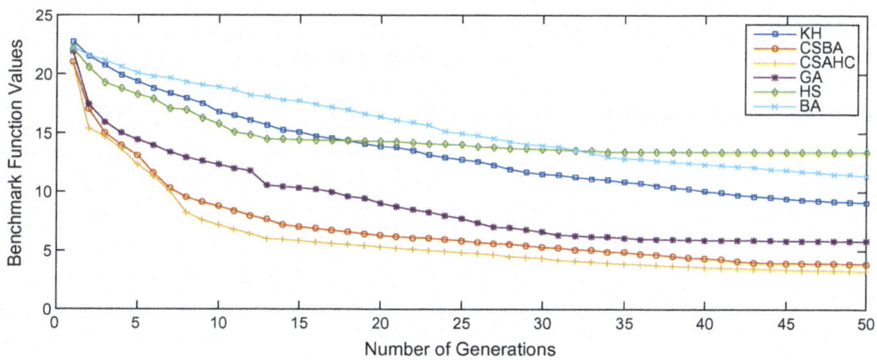

(h) Sphere Function

Fig. 7.9 (continued)

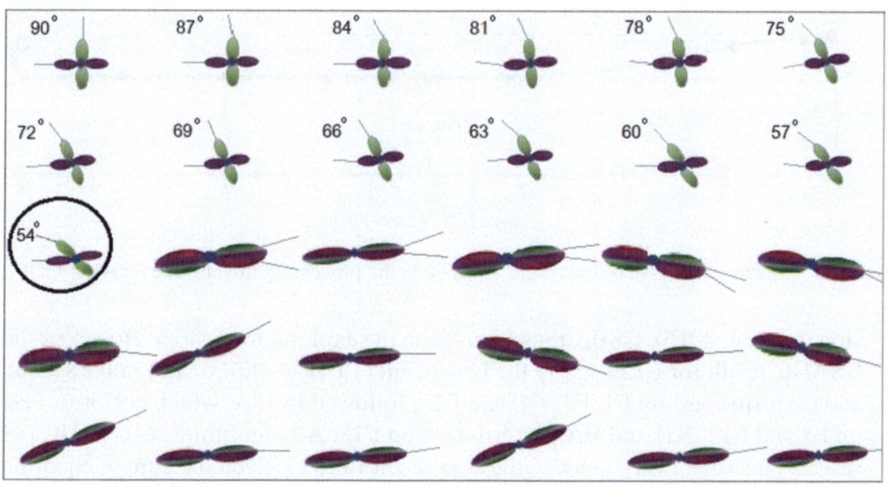

Fig. 7.10 The ability of the CSAHC-ODF to detect maxima separated by 54°

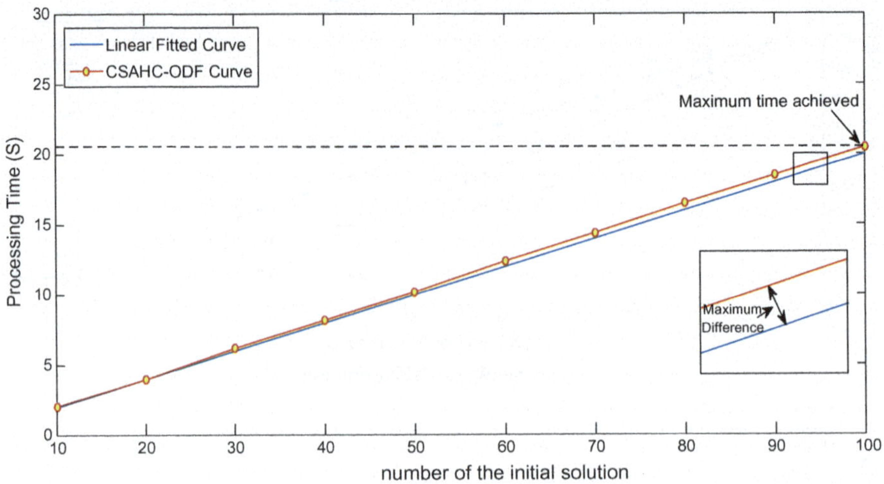

Fig. 7.11 Influence of the number of initial solutions on the processing time for the CSAHC-ODF

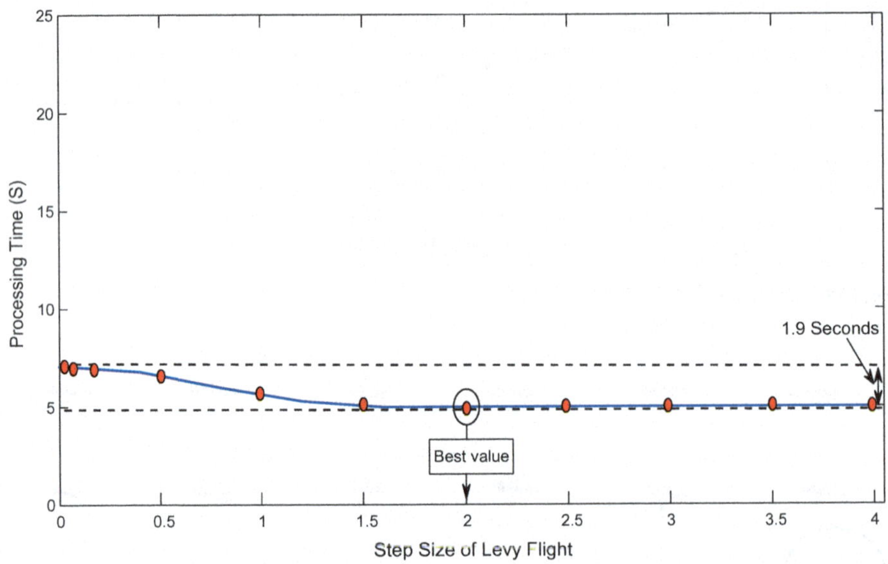

Fig. 7.12 Influence of the step size of *Lévy* Flight on the processing time for the CSAHC-ODF

sion (Li et al. 2016). CSBA thus focuses on the exploitation search. However, the CSAHC result for F7 is nearly the best result (1.01E + 00). CSBA ranks second and performs best on F1, F3, F4, and F13, followed by GA, which performs best on F5 and F13. KH and BA perform best on F13. All algorithms, except HC (its result is near the best), achieve the best result for F13 given the simple equation for this function, which is called *Step*.

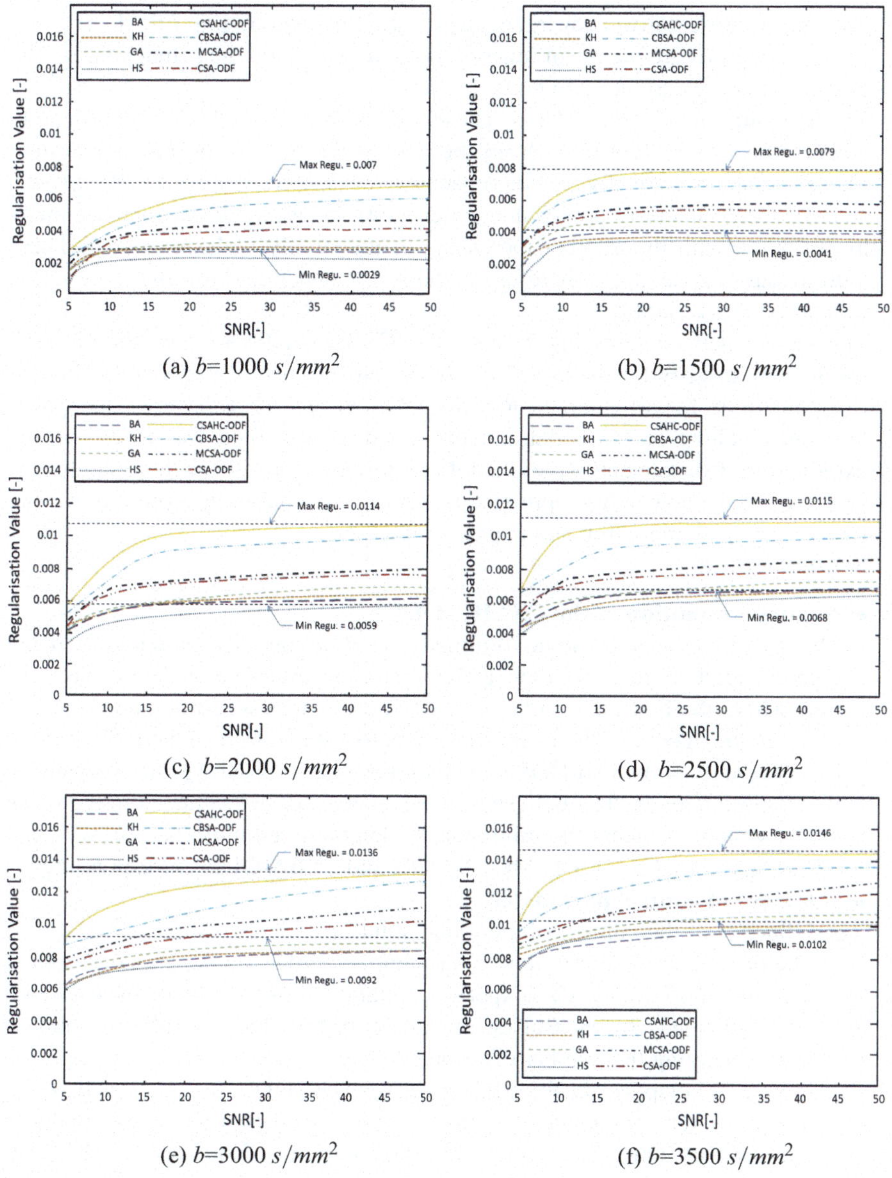

(a) b=1000 s/mm^2 (b) b=1500 s/mm^2

(c) b=2000 s/mm^2 (d) b=2500 s/mm^2

(e) b=3000 s/mm^2 (f) b=3500 s/mm^2

Fig. 7.13 The influence of the b-value on the HS, KH, BA, GA, CSA-ODF, MCSA-ODF, CSBA-ODF, and CSAHC-ODF

The most representative convergent curves are illustrated in Fig. 7.9a–h. The values in the figures are the mean function optima, which are the actual values. The discussion is divided into two parts.
The first part contains Fig. 7.9a–d. These figures show that CSAHC significantly outperforms other algorithms from the first generation to the last generation. These results are logically expected because the *Ackley*, *Penalty #1*, *Penalty #2*, and *Rastrigin* Functions are located under multimodal function, which has many local optima with the number increasing exponentially with dimension. These multimodal functions are therefore suitable for benchmarking the exploration capability of a variant.
The second part contains Fig. 7.9e–h. The CSBA results are near the CSAHC results with priority for the last. This convergence of results is due to the benchmark functions. In particular, the *Rosenbrock*, *Schwefel 1.2*, *Schwefel 2.22*, *Sphere* Functions belong to unimodal functions, which are fitted for benchmarking exploitation. Consequently, these solutions provide a proof for the reliable performance of the new hybrid approach in terms of exploiting the optimum. CSBA focuses on the exploitation and is thus competitive.

2. **Synthetic Data**

- **Angular resolution of the CSAHC-ODF**
This part illustrates the angle minimum to extract the ODF maxima by using CSAHC-ODF. Figure 7.10 depicts the separation angles between the two maxima, where the first ODF in the upper left refers to a separation angle of 90° between the maxima. The next ODF refers to a separation angle of 87°, which is less than 3° of the first ODF and continues to decrease by 3° until acquiring the minimum angle. The CSAHC-ODF achieves 54°, as depicted in the black circle, which considers the minimum angle compared with the 66°, 63°, and 57° achieved by CSA-ODF, MCSA-ODF, and CSBA-ODF, correspondingly.
- **Influence of control parameter**

(a) Influence of the number of the initial solutions
The parameter setting in this experiment is similar to the parameter setting used in Sect. 5.4.2(c). This experiment mainly aims to evaluate the CSAHC-ODF efficiency by measuring its capability to achieve a reliable solution when changing the number of solutions.
Figure 7.11 displays the fitted linear and CSAHC-ODF curves. CSAHC-ODF obtains better time results than those of the fitted linear curve, especially in the first part (until the number of initial solution = 50). The CSAHC-ODF performance is insignificantly affected by increasing the number of solutions and is nearly identical to the fitted linear in the second part. The maximum time achieved when the number of initial solutions equals to 100 is 20.09 s. Also, the best result has been achieved by the difference time between the CSAHC-ODF curve and fitted linear curve which does not exceed 0.1 s. This indicates to the robustness of the CSAHC-ODF by dealing with the large number of initial solutions.

(b) Influence of the *Lévy* Flights
Step size is one of the most important elements in the search process. A decrease in step size increases the search time and vice versa. However, the value of step size should not be very large because it may be out of the search area and lead to missing closely spaced peaks. By contrast, using a small step size may result in slow convergence. Thus, step size should be in between them.
Figure 7.12 illustrates the effect of step size on the CSAHC-ODF performance. The line refers to the processing time in different step sizes. The results show that the best value of step size equals to 2. This result denotes that CSAHC-ODF finds the best solution at the lowest processing time when the step size equals to at least 2.
However, the best value of the step size for CSAHC-ODF equals to 2 according to the problems caused by a large step size. The CSAHC-ODF performance is inconsiderably affected by the increase in step size because the time difference between the lowest and highest values of step size does not exceed 2 s. CSAHC-ODF exhibits search robustness (i.e., exploration and exploitation of the search space).

- **Influence of SNR and *b*-value** :
HS, KH, BA, GA, CSA-ODF, MCSA-ODF, CSBA-ODF, and CSAHC-ODF are compared in this experiment by using different *b*-value (1000, 1500, 2000, 2500, 3000, and 3500) and SNR (5, 10, 15, 20, 25, 30, 35, 40, 45, and 50). The results show the average of 1000 runs for each value.
Figure 7.13 shows the influences of *b*-value and SNR on HS, KH, BA, GA, CSA-ODF, MCSA-ODF, CSBA-ODF, and CSAHC-ODF. The experiments aim to evaluate the performance of each algorithm to find the best regularization value (i.e., the smoothness of the ODF shape is unacceptable). Figure 7.13a–f depict that the regularization value is increased gradually. Figure 7.13a presents the highest regularization value equal to 0.007, which is achieved by CSAHC-ODF. Figure 7.13f indicates the highest regularization value equal to 0.0146, which is also generated by CSAHC-ODF. Most of the results refer to the best SNR value located between 25 and 30, thereby indicating the increase in the regularization value by increasing the *b*-value. The comparison results show that CSAHC-ODF outperforms the other methods. CSBA-ODF is in the second rank, followed by MCSA-ODF, CSA-ODF, GA, KH, BA, and HS.

7.5 Summary of the Comparisons of the CSA-ODF Versions

This section presents the summary of the three versions of the CSA-ODF, where in the previous each version was compared with the versions that preceded it in each chapter. In other words, Chap. 6 contains the comparison between MCSA-ODF

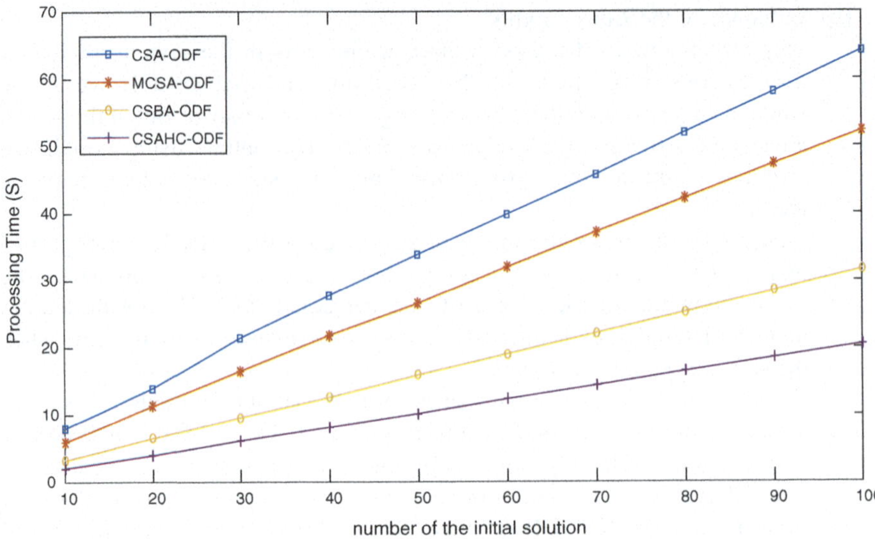

Fig. 7.14 Influence of the number of initial solutions on the processing time for the CSAHC-ODF

and CSA-ODF, Chap. 7 contains comparison between CSBA-ODF and each one of the MCSA-ODF and CSA-ODF, followed by the last comparison in the same chapter which compared CSAHC-ODF, CSBA-ODF, MCSA-ODF, and CSA-ODF). The comparisons focused on three main points, as follows:

- **Angular resolution**:
  In the experiments, each version was used to determine its ability for detecting the ODF maxima with the smallest angle between two fibers. In other words, the minimum angular separation at which two fibers can be separated when at the same range (crossing fibers) that's called angular resolution. Based on the results achieved by each version in this regard, the CSAHC-ODF got the first rank with separation angle 54° followed by CSBA-ODF, MCSA-ODF, and CSA-ODF with 57°, 63°, and 66°, respectively. To prove these results, Table 7.5 shows the success rate for each version in each one of the separation angles. Where it considers the fraction or percentage of success among a number of attempts of each version to

Table 7.5 Success rate (%) in detecting ODF maxima in different separation angle

Algorithm	Separation Angle														
	90° (%)	87° (%)	84° (%)	81° (%)	78° (%)	75° (%)	72° (%)	69° (%)	66° (%)	63° (%)	60° (%)	57° (%)	54° (%)	51° (%)	48°
CSA-ODF	100	100	100	100	98	98	84	84	80	23	2	1	0	0	0
MCSA-ODF	100	100	100	100	100	99	99	91	89	88	18	2	1	0	0
CSBA-ODF	100	100	100	100	100	100	100	100	100	99	98	98	48	3	0
CSAHC-ODF	100	100	100	100	100	100	100	100	100	100	100	99	99	49	4

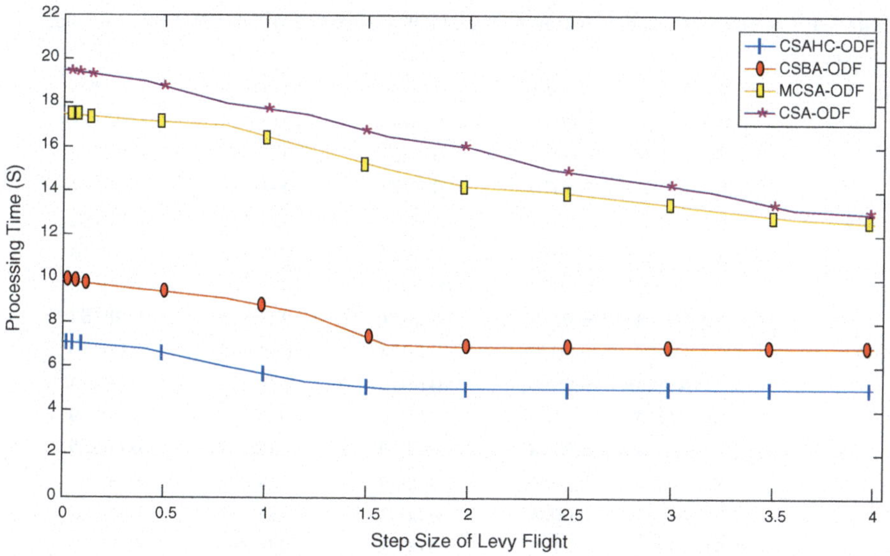

Fig. 7.15 Influence of the step size of *Lévy* Flight on the processing time for the CSA-ODF, MCSA-ODF, CSBA-ODF, and CSAHC-ODF

determine the minimum separation angle during 100 run times with free noise. The results refer to all versions achieved the best success rate when the separation angle between 90° and 81°, because finding the ODF maxima very easy in this phase. However, in the second phase (i.e., 78° until 66°) there are some changing in the results especially for CSA-ODF and MCSA-ODF because the separation angle becomes smaller, so the search process becomes more difficult. At the end, it can be observed the performance of the CSAHC-ODF is more accurate to deal with angular resolution.

It's worth to mention that the higher success rate for the separation angle leads to increase the quality of the results with decreasing the error ratio. Therefore, the minimum success rate percentage shouldn't less than 75% (Cho et al. 2008).

- **Influence of the control parameter**:

This point contains two part: (i)Influence of the number of the initial solutions. (ii) Influence of the step size. Both of them were focused on evaluating the performance of each version in terms of its influenced by these parameters on the processing time. In Fig. 7.14, the results show the effects of increasing the number of initial solution on the processing time for the each algorithm. It can be observed the CSA-ODF and the MCSA-ODF are not stable comparing with CSBA-ODF and CSAHC-ODF, this indicates that there is some weakness in the search method, also that prove the processing time is high comparing to the CSBA-ODF and CSAHC-ODF. Thus, the CSAHC-ODF proved its efficiency comparing with the CSBA-ODF, MCSA-ODF, and CSA-ODF. Where it obtained the least time for all values of the initial solutions. Due to its controlling of the mechanism of the search. Such as, select the best solution that has the highest fitness value through the

Table 7.6 Comparison results between the CSA-ODF versions by different values of *b*-value

b-value	Algorithm	Best	Worst	Mean	Std
1000	CSA-ODF	0.0043	0.0011	0.00340	0.00295
	MCSA-ODF	0.0048	0.0019	0.00423	0.00317
	CSBA-ODF	0.0061	0.0022	0.00587	0.00282
	CSAHC-ODF	**0.0070**	**0.0029**	**0.00541**	**0.00394**
1500	CSA-ODF	0.0057	0.0031	0.00411	0.00182
	MCSA-ODF	0.0059	0.0033	0.00444	0.00118
	CSBA-ODF	0.0070	0.0040	0.00528	0.00130
	CSAHC-ODF	**0.0079**	**0.0041**	**0.00578**	**0.00182**
2000	CSA-ODF	0.0079	0.0034	0.00644	0.00210
	MCSA-ODF	0.0080	0.0046	0.0061	0.00144
	CSBA-ODF	0.0100	0.0052	0.00820	0.00229
	CSAHC-ODF	**0.0114**	**0.0059**	**0.00959**	**0.00259**
2500	CSA-ODF	0.0080	0.0057	0.00096	0.0068
	MCSA-ODF	0.0086	0.0053	0.00781	0.00168
	CSBA-ODF	0.0102	0.0065	0.00864	0.00155
	CSAHC-ODF	**0.0115**	**0.0068**	**0.00900**	**0.00200**
3000	CSA-ODF	0.0108	0.0078	0.00982	0.00149
	MCSA-ODF	0.0113	0.0080	0.00908	0.00192
	CSBA-ODF	0.0123	0.0085	0.01009	0.00212
	CSAHC-ODF	**0.0136**	**0.0092**	**0.00891**	**0.00553**
3500	CSA-ODF	0.0120	0.0091	0.00824	0.00492
	MCSA-ODF	0.0127	0.0088	0.01068	0.00205
	CSBA-ODF	0.0138	0.0097	0.01098	0.00233
	CSAHC-ODF	**0.0146**	**0.0102**	**0.01222**	**0.00186**

MCSA, exploitation search through the CSBA, and local search through CSAHC. It's worth to mention that the results above have been confirmed by calculating the slope for each version, where are the results show as follows: CSA-ODF = 5.62, MCSA-ODF = 4.63, CSBA-ODF = 2.84, and CSAHC-ODF = 0.97. These results prove that the CSAHC-ODF is the best version in this evaluation (i.e., The influence of the initial solutions)

Figure 7.15 illustrates the effect of the step size on the performance of the CSA-ODF, MCSA-ODF, CSBA-ODF, and CSAHC-ODF. The lines refer to the processing time of each algorithm through using different values of the step size. Sometimes, the optimal value of the step size is vary from one algorithm to another, for example, in the CSA-ODF the optimal value of the step size does not stable. In other words, it's obvious that the CSA-ODF heading towards the largest step size which causes some difficulties as mentioned above. Also, the optimal value of the step size for MCSA-ODF does not clear enough, it can be determined between 2 and 2.5 where the processing time is 13.9 s per generation. Because the curve after this values goes down slowly. While the optimal value of the CSBA-ODF and

CSAHC-ODF it can be observed clearly. In other words, 2 is the optimal value of the step size, which it is the first value achieved the least time for both of CSBA-ODF and CSAHC-ODF, 7 and 5 s per generation, respectively. These result prove what mentioned regarding the value of the step size. It worth to mention that the CSAHC-ODF outperform of the others, where the results show the CSAHC-ODF obtained the lowest time in all values of the step size and the lowest difference of time (i.e., the difference between the time which achieved by the smallest and the largest step size) 1.9 s. followed by CSBA-ODF, MCSA-ODF, and CSA-ODF with difference time 3, 5.5, and 7.5 s, repectively.

- **Influence of the *b*-value and SNR**:
  The experiments of the influence of the *b*-value and SNR on the performance of CSA-ODF versions were discussed in the previous section. The outcomes of these experiments referred to the significantly outweigh for the performance of CSA-ODF versions. The evaluation process was based on the regularization or smoothing parameter. Table 7.6 presents the regularization results which obtained by CSA-ODF, MCSA-ODF, CSBA-ODF, and CSAHC-ODF. Also, there are 6 categories of *b*-value (1000, 1500, 2000, 2500, 3000, 3500) with fixed SNR = 30 (The best value of SNR as mentioned in Sect. 7.4.3.2) and 1000 runtimes. The results show the best result, the worst result, the mean, and the standard deviation for each version, the best solution for each experiment is highlighted in bold.
  All results show that the performance of the CSAHC-ODF achieved the best results in all experiments, clearly. In other words, the results prove the accurate performance of the CSAHC-ODF to determine the smooth shape of the ODF, lead to extract the maximum ODF clearly. These are expected results due to the robustness of the search processes where it has the features of the tournament selection (i.e., give the solution with high fitness to be selected), the advantages of BA (i.e., increase the exploitation search with enhancing the convergence), and the mechanism of the HC (i.e., focused on the local search with avoid the local optimum traps).

7.6 Conclusion

This section provides three main points which have been covered in this chapter;(i) Hybridization with population -based algorithms (i.e., hybridization MCSA with component of BA). (ii) hybridization with local search -based algorithm (hybridization CSBA with HC). (iii) Comparative all versions of CSA-ODF together. Regarding the evaluations, same experiments in the previous two chapters have been used in this chapter. In other words, this chapter has been focused on two ways to evaluate the performance of the proposed methods by using benchmark functions then the synthetic data. The following paragraphs includes more detail about the summary of this chapter.

CSBA has the same objective of the MCSA (i.e., solve the slow convergence with enhancing the accuracy of the solution) by focusing on the exploration which considered as a feature of the MCSA and exploitation which considered as a feature of the BA. Therefore the CSBA has the ability to increase the convergence with avoiding premature convergence. CSBA was investigated by applying 13 benchmark functions, the results showed that the CSBA was improved its efficiency of determining the best solution compared to other search algorithms, such as the MCSA, BA, GA, HS, and KH. Then it was applied on the synthetic data, CSBA-ODF showed remarkable superiority compared with other algorithms in all evaluation processes, such as angular resolution, influence of the control parameter, influence of *b*-value, and influence of SNR.

In the hybridization with the local search -based algorithm, CSAHC method is proposed to improve the performance of the CSBA (i.e., improved the quality of the exploitation of CSBA) by integrating it with HC. CSAHC was based on the mechanism of the HC in local search to find the solutions. Therefore, it finds out many local optimum traps by employing both of local and global search.

The same steps were followed to evaluate the performance of the CSAHC on the 13 benchmark functions. The CSAHC improved the quality of the exploration of CSBA by using the HC which is capable of dealing with local searches. The experimental results present that the CSAHC achieved better results compared with the other algorithms. In contrast (i.e., ODF problem), robustness to noise, influence of control parameters, and angular resolution have been used to evaluate the performance of the CSBA-ODF. The CSBA-ODF achieved good results such as it wasn't much affected by increase the noise, it detected the ODF with small angel that's suitable for fiber tracking. Regarding its results, it may selected as a best version.

Finally, all versions of the CSA-ODF were compared to choose the best version for applied it on the real data set. The comparisons covered two points; the time by the influence of the control parameter and accuracy by angular resolution, influence of the *b*-value and SNR. All the results agreed on selected the CSAHC-ODF as the best version between all versions of CSA-ODF. This choice is logically acceptable, because each new version of the CSA-ODF improves the prior version.

References

Abdel-Baset, M., & Hezam, I. M. (2016b). Solving linear least squares problems based on improved cuckoo search algorithm. *Mathematical sciencese*, *5*(2), 199–202.

Abualigah, L. M., Khader, A. T., & Hanandeh, E. S. (2018b). Hybrid clustering analysis using improved krill herd algorithm. *Applied Intelligence*, pp. 1–25.

Abualigah, L. M., Khader, A. T., & Hanandeh, E. S. (2018d). A novel weighting scheme applied to improve the text document clustering techniques. In *Innovative Computing, Optimization and Its Applications*, pp. 305–320. Springer.

Abualigah, L. M., Sawaie, A. M., Khader, A. T., Rashaideh, H., Al-Betar, M. A., & Shehab, M. (2017b). β-hill climbing technique for the text document clustering. *New Trends in Information Technology, 60*.

Alajmi, B. N., Ahmed, K. H., Finney, S. J., & Williams, B. W. (2011). Fuzzy-logic-control approach of a modified hill-climbing method for maximum power point in microgrid standalone photovoltaic system. *IEEE Transactions on Power Electronics*, *26*(4), 1022–1030.

Ardjani, F., Sadouni, K., & Benyettou, M. (2010). Optimization of svm multiclass by particle swarm (pso-svm). In *2010 2nd International Workshop on Database Technology and Applications*, pp. 1–4. IEEE.

Burke, E. K., & Newall, J. P. (2002). Enhancing timetable solutions with local search methods. In *International Conference on the Practice and Theory of Automated Timetabling*, pp. 195–206. Springer.

Cho, K.-H., Yeh, C.-H., Tournier, J.-D., Chao, Y.-P., Chen, J.-H., & Lin, C.-P. (2008). Evaluation of the accuracy and angular resolution of q-ball imaging. *Neuroimage*, *42*(1), 262–271.

Dejam, S., Sadeghzadeh, M., & Mirabedini, S. J. (2012). Combining cuckoo and tabu algorithms for solving quadratic assignment problems. *Journal of Academic and Applied Studies*, *2*(12), 1–8.

Jamil, M., & Yang, X.-S. (2013). A literature survey of benchmark functions for global optimisation problems. *International Journal of Mathematical Modelling and Numerical Optimisation*, *4*(2), 150–194.

Jiao, J., & Long, W. (2014). *Hybrid cuckoo search algorithm based on powell search for constrained engineering design optimization13*, 431–440.

Layeb, A. (2011). A novel quantum inspired cuckoo search for knapsack problems. *International Journal of bio-inspired Computation*, *3*(5), 297–305.

Li, Z., Zhou, Y., Zhang, S., & Song, J. (2016). Lévy-flight moth-flame algorithm for function optimization and engineering design problems. *Mathematical Problems in Engineering*.

Li, X., Wang, J., & Yin, M. (2014). Enhancing the performance of cuckoo search algorithm using orthogonal learning method. *Neural Computing and Applications*, *24*(6), 1233–1247.

Liu, X., & Hui, F. (2014). Pso-based support vector machine with cuckoo search technique for clinical disease diagnoses. *The Scientific World Journal*.

Pavlyukevich, I. (2007). Lévy flights, non-local search and simulated annealing. *Journal of Computational Physics*, *226*(2), 1830–1844.

Reyaz-Ahmed, A., Zhang, Y.-Q., & Harrison, R. W. (2009). Granular decision tree and evolutionary neural svm for protein secondary structure prediction. *International Journal of Computational Intelligence Systems*, *2*(4), 343–352.

Rubio, A., & Gámez, J. A. (2011). Flexible learning of k-dependence bayesian network classifiers. In *Proceedings of the 13th annual conference on Genetic and evolutionary computation*, pp. 1219–1226. ACM.

Schaerf, A., & Meisels, A. (1999). Solving employee timetabling problems by generalized local search. In *Congress of the Italian Association for Artificial Intelligence*, pp. 380–389. Springer.

Shehab, M., & Khader, A. T. (2018). *Modified cuckoo search algorithm using a new selection scheme for unconstrained optimization problems*, *14*, 1.

Shehab, M., Daoud, M. Sh., AlMimi, H. M., Abualigah, L. M., & Khader, A. T. (2019a). Hybridizing cuckoo search algorithm for extracting the odf maxima in spherical harmonic representation. *International Journal of Bio-Inspired Computation*, (in press).

Shehab, M., Khader, A. T., & Al-Betar, M. A. (2016). New selection schemes for particle swarm optimization. *IEEJ Transactions on Electronics, Information and Systems*, *136*(12), 1706–1711. https://doi.org/10.1541/ieejeiss.136.1706.

Shehab, M., Khader, A. T., & Al-Betar, M. A. (2017a). A survey on applications and variants of the cuckoo search algorithm. *Applied Soft Computing*.

Shehab, M., Khader, A. T., & Alia, M. A. (2019b). Enhancing cuckoo search algorithm by using reinforcement learning for constrained engineering optimization problems. In *2019 IEEE Jordan International Joint Conference on Electrical Engineering and Information Technology (JEEIT)*, pp. 812–816. IEEE.

Shehab, M., Khader, A. T., & Laouchedi, M. (2017c). Modified cuckoo search algorithm for solving global optimization problems. In *International Conference of Reliable Information and Communication Technology*, pp. 561–570. Springer.

Shehab, M., Khader, A. T., & Laouchedi, M. (2018a). A hybrid method based on cuckoo search algorithm for global optimization problems. *Journal of ICT*, *17*(3), 469–491.

Shehab, M., Khader, A. T., Al-Betar, M. A., & Abualigah, L. M. (2017b). Hybridizing cuckoo search algorithm with hill climbing for numerical optimization problems. In *Information Technology (ICIT), 2017 8th International Conference on*, pp. 36–43. IEEE.

Shehab, M., Khader, A. T., Laouchedi, M., & Alomari, O. A. (2018b). Hybridizing cuckoo search algorithm with bat algorithm for global numerical optimization. *The Journal of Supercomputing*, 1–28.

Sheikholeslami, R., Zecchin, A. C., Zheng, F., & Talatahari, S. (2016). A hybrid cuckoo–harmony search algorithm for optimal design of water distribution systems. *Journal of Hydroinformatics*, *18*(3), 544–563.

Simon, D. (2008). Biogeography-based optimization. *IEEE Transactions on Evolutionary Computation*, *12*(6), 702–713.

Walton, S., Hassan, O., Morgan, K., & Brown, M. R. (2011b). Modified cuckoo search: a new gradient free optimisation algorithm. *Chaos, Solitons & Fractals*, *44*(9), 710–718.

Wang, G.-G., Gandomi, A. H., Zhao, X., & Chu, H. C. E. (2016b). Hybridizing harmony search algorithm with cuckoo search for global numerical optimization. *Soft Computing*, *20*(1), 273–285.

Yang, X.-S. (2010b). A new metaheuristic bat-inspired algorithm. In *Nature inspired cooperative strategies for optimization (NICSO 2010)*, pp. 65–74. Springer.

Yang, X.-S., & Deb, S. (2013). Multiobjective cuckoo search for design optimization. *Computers & Operations Research*, *40*(6), 1616–1624.

Yang, X.-S., & He, X. (2013). Bat algorithm: Literature review and applications. *International Journal of Bio-Inspired Computation*, *5*(3), 141–149.

Yılmaz, S., & U Küçüksille, E. (2015). A new modification approach on bat algorithm for solving optimization problems. *Applied Soft Computing*, *28*, 259–275.

Chapter 8
Apply CSAHC-ODF Algorithm for Extracting the ODF Maxima from the Human Brain's Data

8.1 Introduction

In this chapter, the CSAHC-ODF is employed to extracting the maximum of the ODF from the real human brain data. CSAHC-ODF has been used because it got the best version (Shehab et al. 2017b, c, 2018a, b) based on the experiments in Sect. 7.5. This makes the CSAHC-ODF is able to deal with the real data which collected from Institut Perubatan dan Pergigian Termaju—Universiti Sains Malaysia, IPPT-USM (Advanced Medical and Dental Institute (AMDI)). The results of the CSAHC-ODF are analyzed using FMRIB Software Library v5.0.8, created by the analysis group, FMRIB, Oxford, UK, which called FSL.

The rest of this chapter is organized as follows. Section 8.2 describes the acquisition of the real data in details. Data reconstructing and extracting the ODF maxima by CSAHC-ODF are presented in Sect. 8.3. The conclusion is illustrated in Sect. 8.4.

8.2 Data Acquisition and Processing

The real human brain's data acquisition from the normal healthy volunteer. The procedures were obtained prior to the experiment according to the guidelines set by IPPT-USM. In order to acquire T2-weighted data, Turbo Spin Echo (TSE) is mainly used (one of the basic pulse sequences in MRI). The imaging pulse sequence was a standard sequence used regularly for clinical studies in the IPPT.

The Digital Imaging and Communications in Medicine (DICOM) dataset was acquired using 60 directions. The diffusion-weighting used was *b*-value = 2000 s/mm^2 and imaging parameters: Spin Echo sequence: TEcho Time (TE) / Repetition Time (TR) = $9000/102/ms$, number of acquisitions = 1, FOV (typically divided into several hundred picture elements (pixels), each approximately 1 mm^2 in size) = 256 mm × 184 mm, matrix size (the number of frequency encoding steps) = 128 × 92, resolution of 2 × 2 mm^2. T2-weighted three-dimensional magnetization-prepared

M. Shehab, *Artificial Intelligence in Diffusion MRI*, Studies in Computational Intelligence 877, https://doi.org/10.1007/978-3-030-36083-2_8

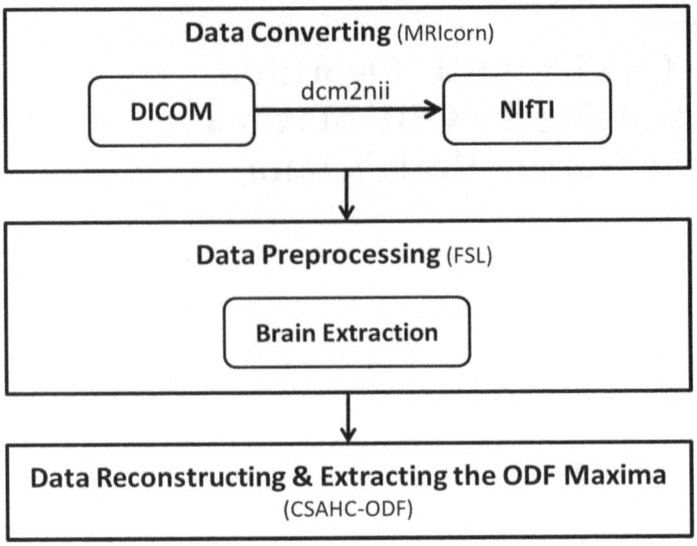

Fig. 8.1 Main procedure for the processing of real datasets

rapid-gradientecho (MP-RAGE) scans of the entire head were also obtained at $1 \times 1 \times 1\ \text{mm}^3$ resolution for anatomical referencing of the volunteer. Acquisition time for the 60 direction diffusion scan was around 15 min (Shehab et al. 2019a).

The workflow for image processing is shown in Fig. 8.1, and followed by a description of each step as follows:

- **Data converting**
  The HARDI data are imported DICOM, which is a standardized medical image format. DICOM provides interoperability across hardware, but was not designed to facilitate efficient data manipulation and image processing. Hence, DICOM need to analysis and image processing. The analysis of neuroimaging data often starts with converting the DICOM data to a data format called Neuroimaging Informatics Technology Initiative (NIfTI), which used to provide coordinated and targeted service, training, and research to speed the development and enhance the utility of informatics tools related to neuroimaging.

  In this research, the converting from DICOM to NIfTI has been done by calling dcm2nii in the MRIcron software.[1] MRIcron it can load multiple layers of images, generate volume renderings and draw volumes of interest, Fig. 8.2 illustrates the interface of the software.
- **Data Preprocessing**
  - *Brain Extraction:* To reduce the memory cost and speed up the processing in subsequent steps, the non-brain space in the raw images such as the neck is

[1] For more info: http://people.cas.sc.edu/rorden/mricron/index.html.

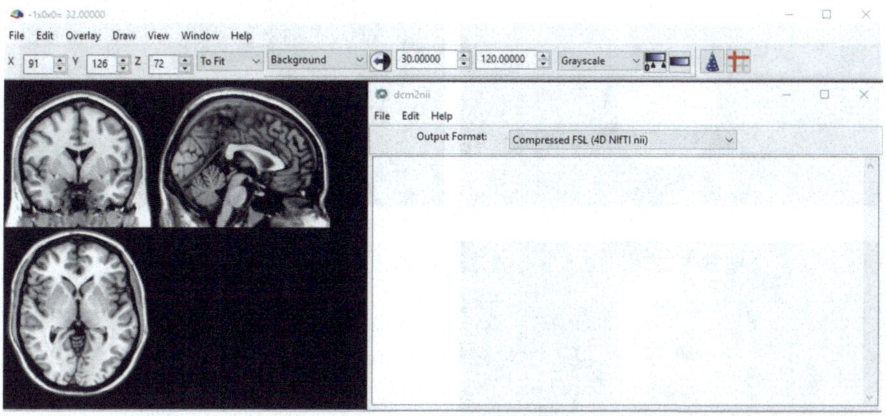

Fig. 8.2 The interface of MRIcron software

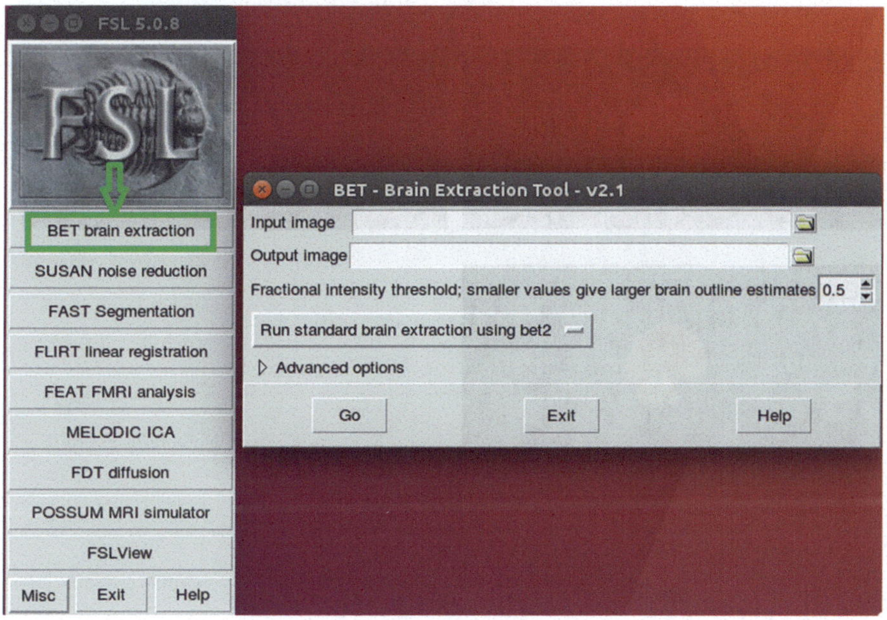

Fig. 8.3 The brain extraction tool of the FSL

cut off, leading to a reduced image size. This step produces the brain mask by using the first command of the FSL, namely, the brain extraction tool (BET), as shown in Fig. 8.3. The brain mask is required for the subsequent processing steps. Figure 8.4 illustrated the original image for the brain and the Image for extracted brain by employing BET.

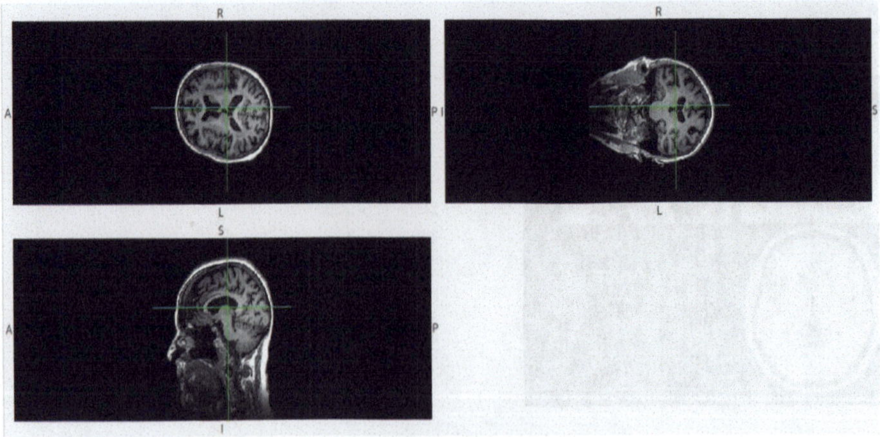

(a) Original Image

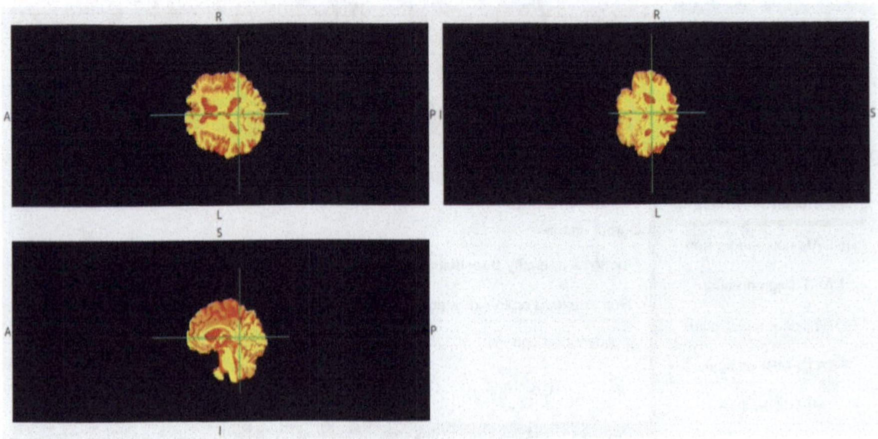

(b) Extracted Brain

Fig. 8.4 Extracting the brain shape from the whole image

8.3 Data Reconstructing and Extracting the ODF Maxima Through CSAHC-ODF

In this research, reconstruction the QBI ODFs estimated using the method described in (Descoteaux et al. 2007; Shehab et al. 2017a), where it requires spherical interpolation and employs spherical radial basis functions as mentioned in Sect. 4.3.2. The QBI signal was expressed as a SH series of order L = 4 and the Funk-Radon transform. The final ODF reconstruction was shown in Eq. 4.1. The aim is to recon-

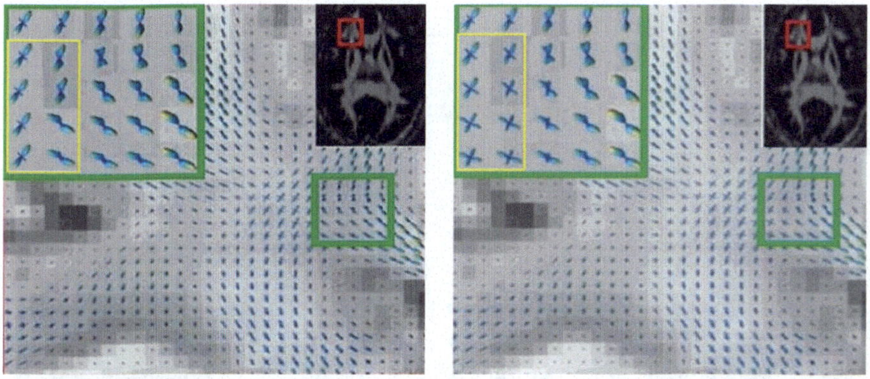

(a) Reconstruction of the QBI ODFs

(b) Reconstruction of the SH QBI ODFs with CSA

Fig. 8.5 The diffusion-weighted data in the brain(vivo), region-of-interest (ROI) outlined in black in the parietal lobe where projections from the corpus callosum (CC). **a** shows the reconstruction of the QBI ODFs. In the same ROI, **b** presents the reconstruction of the spherical harmonic QBI ODFs with CSBA

struct the diffusion strength at each voxel and in each direction. Hence, the unknown orientation distribution function depends on two angular variables (θ, ϕ).

Figure 8.5 illustrates the reconstructed ODFs, where the Fig. 8.5a refers to the reconstruction the ODF from the QBI and Fig. 8.5b refers to the reconstruction the ODF from spherical harmonic QBI through the CSAHC-ODF. Regarding the ROI, the corpus callosum (CC) has been chosen, that connects neurons in the left and right cerebral hemispheres. It considers as the region of the largest fiber tract in the brain (Aboitiz and Montiel 2003; Björnholm et al. 2017).

All results of the reconstruction ODFs are determined without min-max normalization (i.e., ODFs are shown as they are). CSAHC-ODF produces sharper and more accurate ODFs than the original QBI (see Sect. 7.4.3(b)-2). In the green rectangle in the upper left of the figure, can be observed the difference of the ODFs shapes between the original QBI and the reconstructed QBI by CSAHC-ODF. In addition, although sharpening enhances the original QBI ODFs considerably in anisotropic tissue, it causes significant instability in isotropic regions. In other words, it reconstructs the ODF irregular manner. In contrast to CSAHC-ODF which preserves isotropy fairly well, due to the principle of CSAHC-ODF where it looking for all points in the SH to reconstruct the ODF.

The last step in this chapter is extracting the maxima of the ODF from the human brain data. After CSAHC-ODFs were reconstructed in the fourth order RSSH basis from 60 diffusion images acquired at $b - value = 2000\ \mathrm{s/mm}^2$. Figure 8.6 shows the ODFs with their extracted maxima superimposed on the fractional anisotropy (FA) map, in the region of the fiber bundles intersect (CC). To prove the performance of the CSBA-ODF, all the maxima are shown in the zoomed zone, including those corresponding to slight variations in the ODF, such as due to noise. Nevertheless,

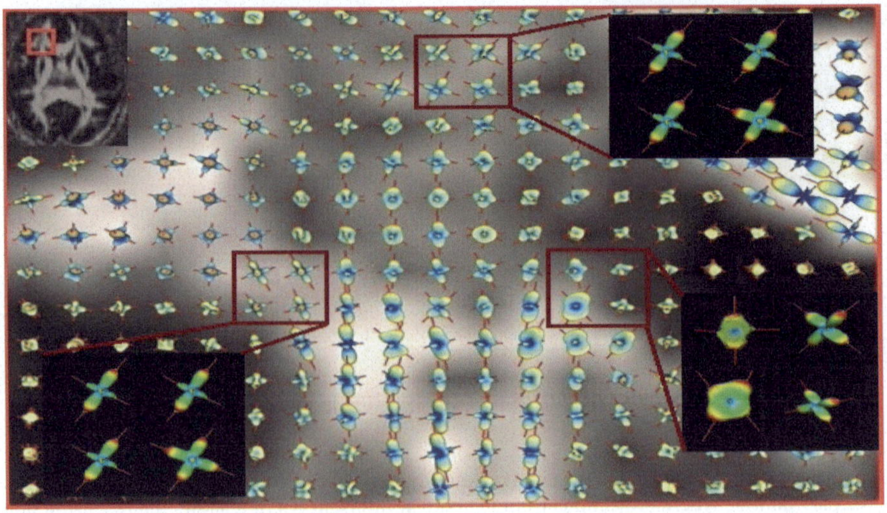

Fig. 8.6 The maxima extraction on human brain data

the prime ODF peaks corresponding to fiber orientations were selected by placing a threshold on the ODF or on its curvature (i.e., the most complicated points).

8.4 Conclusion

In this chapter, the real data of the human brain was used to evaluate the CSAHC-ODF. In the beginning, the data was processed through the MRIcron software which converted the data from DICOM to NIfTI, followed by the data preprocessing where it handled by using the FSL software. Finally, data reconstructing and extracting the ODF maxima through CSAHC-ODF. In the experiments of the reconstruction, the results show the CSAHC-ODF achieved clearer and more accurate ODFs comparing with original QBI. In addition, despite the presence of noise, the results show that all ODF maxima were detecting clearly in the CC which considered as the most complicated area in the brain. The results of this the CSAHC-ODF provide the full maps of the ODF maxima to start the tractography processes. It's worth to mention that the medical side was done under direct guidance from Dr. Makhlouf Laouchedi $ParIMd$, LRPE (USTHB) and $HpitalCentraldelArme$, Alger. Also, some assistance from Iman Ajang; Assistant Professor of Department of Radiology, Massachusetts General Hospital/Harvard Medical School, Charlestown.

References

Aboitiz, F., & Montiel, J. (2003). One hundred million years of interhemispheric communication: The history of the corpus callosum. *Brazilian Journal of Medical and Biological Research, 36*(4), 409–420.

Björnholm, L., Nikkinen, J., Kiviniemi, V., Nordström, T., Niemelä, S., Drakesmith, M., et al. (2017). Structural properties of the human corpus callosum: Multimodal assessment and sex differences. *NeuroImage, 152*, 108–118.

Descoteaux, Maxime, Angelino, Elaine, Fitzgibbons, Shaun, & Deriche, Rachid. (2007). Regularized, fast, and robust analytical Q-ball imaging. *Magnetic Resonance in Medicine, 58*(3), 497–510.

Shehab, M., Khader, A. T., & Al-Betar, M. A. (2017a). A survey on applications and variants of the cuckoo search algorithm. *Applied Soft Computing*.

Shehab, M., Khader, A. T., Al-Betar, M. A., & Abualigah, L. M. (2017b). Hybridizing cuckoo search algorithm with hill climbing for numerical optimization problems. In *2017 8th International Conference on Information Technology (ICIT)* (pp. 36–43). IEEE.

Shehab, M., Khader, A. T., & Laouchedi, M. (2017c). Modified cuckoo search algorithm for solving global optimization problems. In *International Conference of Reliable Information and Communication Technology* (pp. 561–570). Springer.

Shehab, M., Khader, A. T., & Laouchedi, M. (2018a). A hybrid method based on cuckoo search algorithm for global optimization problems. *Journal of ICT, 17* (3), 469–491.

Shehab, M., Khader, A. T., Laouchedi, M., & Alomari, O. A. (2018b). Hybridizing cuckoo search algorithm with bat algorithm for global numerical optimization. *The Journal of Supercomputing*, 1–28.

Shehab, M., Daoud, M. S., AlMimi, H. M., Abualigah, L. M., & Khader, A. T. (2019a). Hybridizing cuckoo search algorithm for extracting the ODF maxima in spherical harmonic representation. *International Journal of Bio-Inspired Computation*. in press.

Chapter 9
Conclusion and Future Work

9.1 Introduction

This chapter provides a summary of the research work and contributions against the research objectives together with the conclusions as to what has been achieved. As well as recommending some future work directions for those interested in the field.

9.2 Summary of Contributions

This research has provided alternative approaches to solve the Orientation Distribution Function Problem (ODF) using Cuckoo Search Algorithm (CSA). The first contribution has been made in the algorithm that was called Basic Cuckoo Search Algorithm (BCSA) which includes the mathematical formulation of the constructing the ODF. The scope of this method is limited to the suitable search space areas. Thus, feasibility is preserved using three ways; benchmark functions, synthetic data, and real data (just for the best version), respectively. These ways incorporated to evaluate the performance of the proposed versions. Through the search, the operators were adjusted to guarantee the feasibility at each iteration. Then, a comprehensive analysis of the convergence behavior was behaved. The results proved promising despite two weaknesses: (i) the slow convergence (ii) weak local exploitation capability.

Special enhancements were made to the BCSA leading to three versions (i.e., MCSA, CSBA, and CSAHC). These versions are used for addressing a particular weakness of the BCSA. For example, provide a targeted rather than a random selection mechanism and thus improve the quality of the chosen solution, this considered as a preface to overcome the first weakness. Also, combining with other algorithms which focus on the local exploitation weakness. In MCSA, the random selection scheme in the BCSA is replaced with tournament selection scheme to increase the fitness of results which lead to elaboration the speed of convergence. Regarding the CSBA, the benefits of the BA is used By integrating some of its components with

M. Shehab, *Artificial Intelligence in Diffusion MRI*, Studies in Computational Intelligence 877, https://doi.org/10.1007/978-3-030-36083-2_9

MCSA to enhance the search efficiency by focusing on the exploitation ability. Therefore, overcome the weaknesses of the slow convergence. Finally, in CSAHC, the hill climbing optimizer is proposed as a new operator working to improve the search mechanism of the CSBA, By limiting the search space to local optimal solutions.

It is worth mentioning that after ensuring the validity of each version of BCSA (MCSA, CSBA, and CSAHC) through the set of benchmark functions and comparing with different algorithms (GA, KH, HS, BA, and BCSA), each version was employed to solve the ODF problem, namely, CSA-ODF, MCSA-ODF, CSBA-ODF, and CSAHC-ODF. The synthetic data was used to evaluate the performance each version. After that, the best version (i.e., CSAHC-ODF) was applied on the real data set.

9.3 Contributions Against Objectives

The first objective was to adapt the BCSA for extracting the maxima of ODF and it was achieved by CSA-ODF. Though the results were promising, two weaknesses were identified.

The second objective was to enhance the adapted BCSA (i.e., CSA-ODF) through modifying the selection mechanism of the BCSA and incorporating two hybridization techniques of metaheuristic components to overcome the weaknesses of CSA-ODF. (i) The modifying of BCSA called MCSA, to achieve this part, the tournament selection scheme was replaced with the random selection scheme to increase the opportunity for the solution with high fitness value to be selected (Shehab et al. 2017a). Also, approve the speed of convergence. Later, the MCSA was used to solve the ODF problem to be called MCSA-ODF. (ii) The first hybridization called CSBA, that merging the MCSA with some components of the BA to enhance the shortcomings of the exploitation search (Shehab et al. 2017b, 2018b). Then, this version was employed on the ODF problem, namely, CSBA-ODF. (iii) The second hybridization called CSAHC, incorporates the CSBA with local search—based algorithm (i.e., HC) to provide more structured local search and thus enhance exploitation (Shehab et al. 2018a). This version called CSAHC-ODF which used to solve ODF problem.

The third objective was to apply the best version of the CSA-ODF on the human brain's data. To achieve this objective, the CSA-ODF, MCSA-ODF, CSBA-ODF, and CSAHC-ODF were compared together. The CSAHC-ODF achieved the best results comparing with others and it was applied on the real data.

9.4 Future Research

The research has demonstrated the applicability of CSA for extracting the maxima of the ODF. The possible future research directions can be suggested as follows:

- Extend the work of extracting ODF (i.e., CSAHC-ODF) to cover the white matter fiber tracking.

- Investigate a new methods for more enhancing of the angular resolution which increases the accuracy of images.
- Investigate the effect of the selection mechanisms on the performance of the BCSA, such as roulette-wheel selection, linear rank selection, exponential rank selection, steady-state selection and some others
- Other means of hybridization of the BCSA can be investigated such as the simulated annealing and tabu search to escape local optima. Furthermore, the swarm-based algorithm such as particle swarm optimization, artificial bee colony.
- Investigate the effect of the Graphics Processing Unit (GPU) on the performance of the CSAHC-ODF to reduce computation time.
- The proposed versions of BCSA can be applied in real problems. Based on the results which have been achieved compared with other algorithms, these versions may suitable for different real problems in various fields. Such as, power energy savings, engineering, image processing, and clustering and data mining.

References

Shehab, M., Khader, A. T., & Laouchedi, M. (2017a). Modified cuckoo search algorithm for solving global optimization problems. In *International Conference of Reliable Information and Communication Technology*, 561–570. Springer.

Shehab, M., Khader, A. T., Al-Betar, M. A., & Abualigah, L. M. (2017b). Hybridizing cuckoo search algorithm with hill climbing for numerical optimization problems. In *2017 8th International Conference on Information Technology (ICIT)* (pp. 36–43). IEEE.

Shehab, M., Khader, A. T., & Laouchedi, M. (2018a). A hybrid method based on cuckoo search algorithm for global optimization problems. *Journal of ICT*, *17*(3), 469–491.

Shehab, M., Khader, A. T., Laouchedi, M., & Alomari, O. A. (2018b). Hybridizing cuckoo search algorithm with bat algorithm for global numerical optimization. *The Journal of Supercomputing*, 1–28.

The manufacturer's authorised representative in the EU is Springer
Nature Customer Service Centre GmbH, Europaplatz 3, 69115 Heidelberg,
Germany. If you have any concerns regarding our products, please
contact ProductSafety@springernature.com

Printed and bound by CPI Group (UK) Ltd, Croydon, CR0 4YY
25/09/2026
02222892-0019